Human Origins

Human Origins

7 million years and counting

NEW SCIENTIST

New
Scientist

First published in Great Britain by John Murray Learning in 2018
An imprint of John Murray Press
A division of Hodder & Stoughton Ltd,
An Hachette UK company

This paperback edition published in 2022

1

B format ISBN 978 1 529 38201 3
eBook ISBN 978 1 473 62981 3

Typeset by KnowledgeWorks Global Ltd.

Printed and bound in Great Britain by Clays Ltd, Elcograf S.p.A.

John Murray Press policy is to use papers that are natural, renewable and recyclable
products and made from wood grown in sustainable forests. The logging and
manufacturing processes are expected to conform to the environmental regulations of
the country of origin.

John Murray Press
Carmelite House
50 Victoria Embankment
London EC4Y 0DZ

Nicholas Brealey Publishing
Hachette Book Group
Market Place, Center 53, State Street
Boston, MA 02109, USA

instantexpert.johnmurraylearning.com

Contents

Series introduction

New Scientist's *Instant Expert* books shine light on the subjects that we all wish we knew more about: topics that challenge, engage enquiring minds and open up a deeper understanding of the world around us. *Instant Expert* books are definitive and accessible entry points for curious readers who want to know how things work and why. Look out for the other titles in the series:

Contributors

Series Editor: Alison George, *New Scientist*

Editor: Michael Marshall, freelance science journalist based in Devon, UK

Instant Expert Series Editor Jeremy Webb, *New Scientist*

Guest contributors

Justin L. Barrett is a professor of psychology and chief project developer for Fuller Theological Seminary's Office for Science, Theology, and Religion Initiatives. He is the author of *Born Believers: The Science of Children's Religious Belief* (2012). He writes about our natural receptivity to religious ideas in Chapter 7.

Michael Bawaya is a the editor of *American Archaeology* magazine. He lives in Albuquerque, New Mexico and writes in Chapter 6 about human migration into the Americas.

David Begun is a professor of anthropology at the University of Toronto, Canada. He is author of *The Real Planet of the Apes* (2015) and writes here about the dawn of the apes in Europe in Chapter 1.

Ara Norenzayan is a professor of psychology at the University of British Columbia in Vancouver, Canada. He is the author of *Big Gods: How Religion Transformed Cooperation and Conflict* (2013). In Chapter 7 he discusses religion's critical role in helping societies grow.

Mark Pagel is a professor of evolutionary biology at Reading University, UK, a fellow of the Royal Society, and author of *Wired for Culture: Origins of the Human Social Mind* (2012). In Chapter 8 he describes why teaching is critical to our cultural evolution, and asks why language evolved.

Pat Shipman is a palaeoanthropologist and retired Adjunct Professor of Biological Anthropology at Pennsylvania State University. Her latest book is *The Invaders: How Humans and Their Dogs Drove Neanderthals to Extinction* (2015). Her work on the relationship between humans and wolves appears in Chapter 5.

Thanks also to the following writers and editors:

Robert Adler, Claire Ainsworth, Shanta Barley, Colin Barras, Emily Benson, Alyssa A. Botelho, Catherine Brahic, Ewen Callaway, Andy Coghlan, Kate Douglas, Alison George, Jessica Hamzelou, Jeff Hecht, Bob Holmes, David Holzman, Jude Isabella, Ferris Jabr, Dan Jones, Alice Klein, Will Knight, Roger Lewin, Mairi Macleod, Phil McKenna, Rachel Nowak, Jan Piotrowski, James Randerson, Kate Ravilious, David Robson, Michael Slezak, Laura Spinney, Mićo Tatalović, Jeremy Webb, Clare Wilson, Sam Wong, Aylin Woodward, Ed Yong, Emma Young.

Introduction

There can be no more profound question than 'Where do we come from?' Since our ancestors figured out how to think, we humans have wondered about our origins, but it is only within the last few centuries that we have tackled the question scientifically.

The story of the study of human evolution is an epic one, full of extraordinary discoveries, daring adventures and (often) spectacularly vehement arguments. It encompasses a swathe of technical disciplines, and forces us to ask deep questions about who we are as a species. This book is your introduction to the subject.

The first seven chapters tell the story of human evolution in chronological order, beginning with the earliest primates, moving through the earliest ape-like hominins, and concluding with the rise of modern civilization. The final three chapters then step back to ask the three biggest questions: what is so special about us, how did we get that way, and what do we still not know?

That last point is a crucial one. We should admit in advance, with apologies, that no reader will get to the end of this book and find that they understand how humanity evolved. Our understanding of human evolution has been upended, or at least seriously complicated, by a swathe of remarkable discoveries made since the year 2000. So you won't find the ultimate truth here, but you will find plenty of facts, our best explanations for them and, we hope, the right questions to ask.

Michael Marshall, Editor

I
Distant origins

The story of our origins begins tens of millions of years ago. In the wake of a mass extinction that wiped out all the dinosaurs – except for birds – and a host of other creatures, a new group of animals arose. They were small and probably seemed utterly insignificant at first. But they would spread to every continent and ultimately change the face of the planet. They were the primates.

The story of primate evolution spans 55 million years and hundreds of species. But from the point of view of our own evolution, there are four key steps:

1 *The original primates*
2 *The 'higher primates' or 'simians' – the group that includes monkeys*
3 *The apes, especially chimpanzees*
4 *The rise of hominins like us.*

In this chapter we will focus on the first three steps. The following six chapters will cover the hominins.

Meet Archie

Our distant ancestors most likely evolved in Asia, in a hothouse world newly free of dinosaurs. More than 55 million years ago, in the lush rainforests of what is now east Asia, a new voice was heard in the animal chorus: the cry of the first primate.

A fossil unveiled in 2013 might give us an idea of what this crucial ancestor looked like. *Archicebus achilles* is the earliest primate skeleton ever found (see Figure 1.1). It also strongly suggests that our lineage evolved in Asia, several million years earlier

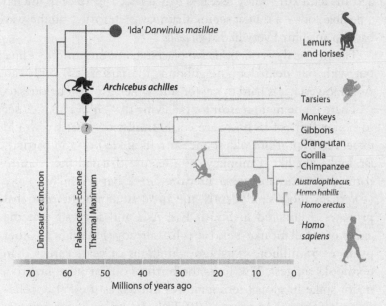

FIGURE 1.1 The evolution of the primates, apes and hominins: *Archicebus achilles* is the oldest primate found so far. Analysis of the fossil places it in the tarsier lineage but it may yet turn out to be a human ancestor.

than we thought, and links the evolution of primates to the most extreme episode of climate change of the last 65 million years.

The American palaeontologist Christopher Beard and his colleagues found *Archicebus* in eastern China, just south of the Yangtze River. It dates from 55 million years ago, has the relatively small eyes of an animal active during the day and the sharp molar teeth of an insect-eater. Significantly, it also has the hind limbs and flexible foot of a primate that had already taken to leaping between branches and gripping them with its feet – characteristics that we lost only when our ancestors left the trees just a few million years ago. In fact, a 2013 study revealed that at least 1 in 13 of us still has a flexible foot – a trait, it seems, that may be traceable all the way back to an animal very like *Archicebus*.

The first analysis of *Archicebus* placed it, not on our direct line, but with our next-door neighbours, the tarsiers of south-east Asia. However, it is hard to say for sure which group it belongs to.

There is one main reason for believing that *Archicebus* is closer to home. Parts of its body are eerily similar to what we would expect to find in our oldest ancestor. Its ankle bone, in particular, looks just like a monkey's – a feature that led the team to name the remarkable fossil after the Greek hero Achilles.

Perhaps most significantly, the fossil supports the idea that primates originated in South East Asia and suggests that the ancestors of all monkeys and apes had already split off from other primates 55 million years ago – millions of years earlier than textbooks suggest. This links the birth of our primate line to a major spike in global temperatures known as the Palaeocene-Eocene Thermal Maximum (PETM). It also puts our point of origin squarely in the heart of the PETM furnace: equatorial Asia. Primates are well adapted to the tropics, so it makes sense that they originated in a warm climate.

What's more, South East Asia would have offered a refuge for tropical species to weather the storm of cooler times: while the Earth's drifting tectonic plates dragged all major continents across the latitudes, this region remained where it was, right on the equator. Many ecological niches were open for the taking after the great mass extinction, and the hot climate would have produced plenty of insects and fruit.

However, this leaves us with a puzzle. As we will see in later chapters, Africa was the cradle of humanity, so at some point early monkey-like primates must have moved from Asia over to Africa: perhaps sometime around 40 million years ago. This is hard to explain, because at that time the vast Tethys Sea separated Asia and Africa.

Monkey archaeology

If you trace the family history of primates, monkeys are somewhere in the middle: further along than the oldest groups like lemurs but not as human-like as the apes. Yet they have achieved something truly remarkable.

In 2016 the first 'monkey archaeology' dig uncovered the tools used by previous generations of wild macaques – a group of primates separated from humans by some 25 million years of evolution. The discovery means that humans aren't alone in leaving a record of past culture that can be pried open by archaeology.

All sorts of animals can use tools, but they are usually made from perishable materials like leaves and twigs. This makes the origin of this behaviour difficult to study, but Burmese long-tailed macaques are a rare exception. They are renowned for their use of stone tools to crack open shellfish, crabs and nuts, making them one of the very few primates that have followed hominins into the Stone Age.

Michael Haslam at the University of Oxford and his team conducted the dig on the small island of Piak Nam Yai in Thailand, one of the islands where the monkeys live and use stone tools. They sifted through the sandy sediments at the site and found ten stone tools attributed to macaques, based on their wear patterns. By dating the oyster shells found in the same sediment layers, they determined that the tools could be as old as 65 years, going back two macaque generations.

We know from eyewitness accounts that these monkeys have been using tools for at least 120 years, so the study doesn't push the age of the behaviour back. But Haslam sees it as a first step towards digging deeper into the origins of the behaviour.

Exactly how far back in time the macaques' Stone Age extends is anyone's guess. A rare 'chimpanzee archaeology' dig in the early 2000s showed that chimps have been using stone tools for more than 4,000 years.

The dawn of the apes

Today, most apes are relatively rare and isolated. Chimpanzees (see Figure 1.2) and gorillas are confined to a few small patches of Africa, while orang-utans are found only on Borneo and Sumatra. Only gibbons roam more widely. Yet if you could look back in time to between 20 and 7 million years ago, science fiction was science fact: the Earth really was the planet of the apes.

At least a hundred species roamed the world before the first humans appeared. They were remarkable in number and diversity, but are more fascinating still for what they tell us about our own origins. Key human traits – including big brains, dexterous

FIGURE 1.2 Apes may have originated in Africa but critical changes took place among European species.

hands, erect posture and long childhood – can be traced back to this period. And the really surprising thing is that these features all evolved in European apes.

There is no doubt that apes originated in Africa, or that our more recent evolution happened there. But for a time between these two landmarks, apes hovered on the verge of extinction on their home continent while flourishing in Europe. What's more, the transformation of European species during this time made us who we are.

The fossil record indicates that apes started out in Africa about 26 million years ago, and were firmly settled there 4 million years or so later in the form of *Proconsul*. A close relative of *Proconsul* named *Ekembo*, which lived 18 million years ago, paints an extraordinary picture of these early apes.

Remains of *Ekembo* were found preserved Pompeii-like in layers of volcanic ash on Rusinga Island, Kenya. What we have is an animal with arms and legs of equal length, a horizontally oriented backbone and a brain about the size of a modern baboon's. In other words, *Ekembo* looks like a largish monkey, but with a key difference: no tail. Tails allow many monkeys to balance, but *Ekembo* compensated with more limber wrists and hips and more powerful hands and feet for grasping. This set apes on a different path from Old World monkeys.

A second momentous change arose in a contemporary of *Ekembo* called *Afropithecus*. The two look remarkably similar from the neck down, but have quite different jaws and teeth. Those of *Afropithecus* are far more robust, adapted for powerful crushing and grinding. Equipped like this, it could extract nutrients from foods with husks and shells impenetrable to the more slender jaws of *Ekembo*. It may not sound too impressive, but this ability had huge repercussions for apes. With the capacity to eat a wider variety of foods, they could expand their range out of Africa and into Europe and Asia.

The oldest apes we know of in Europe belong to the genus *Griphopithecus* and date from 17.5 million years ago. They inherited the powerful bite of *Afropithecus* but their teeth were a little different, more like those of our earliest direct ancestors in Africa. According to the fossil record, griphopiths were living in parts of what are now Germany and Turkey, about 17 million years ago. At this time much of Europe was in a subtropical zone. Seasonality was low and the climate was suitable for animals, like apes, that rely on a continuous year-round supply of fruit. However, as griphopiths migrated north, conditions would have proved more challenging, ultimately driving them to evolve new adaptations.

As well as moving northwards, griphopiths returned south, so that by some 15 million years ago their range covered an area from Europe to East Africa. One member of the family, *Nacholapithecus*, living in Kenya at around this time, had evolved limbs with larger elbows and wrists, perhaps anticipating the development of the longer arms found in living apes and the earliest humans. However, griphopiths seem to die out in Africa, though we don't know why. In fact, the fossil record indicates that between 14 and 8 million years ago apes were a rarity there and most were from ancient lineages related to *Ekembo* and bound for extinction.

By contrast, in Europe, truly modern-looking great apes were emerging. Around 12.5 million years ago, the first ape with a more upright posture appeared. *Pierolapithecus*, sometimes called *Dryopithecus*, was unearthed in Catalonia, northern Spain. The partial skeleton has a more vertical backbone, a broad chest, arms longer than legs, very mobile wrists, and long, curved, powerfully grasping fingers. These features made *Dryopithecus* look more like today's great apes. They also indicate a major transition from walking like a monkey on all fours to ape-like movement, hanging and swinging below branches.

Hispanopithecus, living in what is now Catalonia a few million years later, had longer arms and an even more upright back. So did *Rudapithecus*, its contemporary in what is now Hungary. More significantly, to our knowledge, *Rudapithecus* is the first ape to evolve two other key features of modern great apes: a big brain and extended childhood. In 1999 scientists recovered a well-preserved *Rudapithecus* skull from the site at Rudabánya. Structural details – including the braincase, jaw and base of the skull – all resemble the anatomy of living African apes, especially gorillas, only smaller. The brain was comparable in size to that of living chimpanzees. And evidence from dental growth studies indicates that *Rudapithecus* had a longer childhood than its ancestors.

The anthropologist David Begun has argued that these and other key developments in ape evolution were stimulated by the challenging ecological conditions apes encountered in Europe. Apes colonized the continent during the warmest phase of the Miocene, but by 14 million years ago it was cooling, and forests were becoming less dense and food scarcer. To survive, apes had to develop new strategies to find food both in the trees and on the ground. This led to physical and cognitive changes. Big brains and extended childhoods are associated with higher levels of intelligence, memory, complex learning and strategic thinking, important tools for apes living in challenging seasonal environments – and characteristic attributes of our own species.

Gradually, conditions in Europe became too tough for apes and about 10 million years ago they quit the continent for Africa. There, the separate lines of our closest living relatives evolved, the gorillas branching off first and then chimps and humans veering apart. But the anatomy and behaviour of the earliest humans make sense only in the light of the Miocene apes. It may be that, without the developments that happened in Europe, humans would never have evolved.

The missing link

What about the last common ancestor of humans and chimps? Despite dozens of new fossils being found every year, the original 'missing link' remains as elusive as ever. On the face of it, there is good reason to think that the last creature from which both humans and chimps – our closest cousins – can claim descent might eventually be found. After all, we have a pretty good idea when and where it was dragging its knuckles or swinging through the trees. Most palaeontologists have tended to accept that the last common ancestor of chimps and humans lived in Africa, probably around 7 million years ago.

The bad news is that any evidence of this animal will be very hard to find. After decades of searching, we have a reasonably rich collection of fossils of our hominin ancestors, stretching back 4 million years. But fossil evidence of anything earlier than that would barely fill a couple of shoe boxes. This is partly because hominins lived in places where animals are more likely to fossilize, like lake shores and caves. Earlier relatives may not have done so. We might not even know the remains of the human–chimp ancestor if we saw them. Different scholars have very different expectations of how this ancestor looked.

By comparing early hominin fossils, ape fossils and large numbers of living primates, Sergio Almécija of George Washington University in Washington, DC has concluded that our forebear had hands and thigh bones that were more human-like than chimp-like. It probably still walked on all fours, he says, but not in the way that chimps do. Nathan Young at the University of California, San Francisco and his colleagues have used a broadly similar approach to suggest that the animal's shoulders were chimp-like – suggesting that it swung through the trees as chimps do today. Almécija thinks it is possible that this ancestor had a combination of features – and maybe even some seen in neither group today. One hope is that comparing the genomes of living apes might provide evidence that everyone can agree on.

There is a big caveat, however. All this assumes that there was a single ancestor. Genetic studies so far hint that some of our chromosomes diverged from the chimp versions much earlier than others, possibly indicating that there wasn't a simple, clear split. Rather, primate-like populations separated for a time, then came back together and hybridized before splitting permanently – all over the course of millions of years. Try picking a single ancestor out of that tangled mess.

And there is another problem: we may have been looking at the wrong period of prehistory altogether.

Our true dawn

Line them up in your head. Generation after generation of your ancestors, reaching back in time through civilizations, ice ages, an epic migration out of Africa, to the very origin of our species. And on the other side, take a chimp and line up its ancestors. How far back do you have to go, how many generations have to pass, before the two lines meet?

New estimates for when our lineage and chimps went their separate ways suggest that some of our established ideas are staggeringly wrong. If these suggestions are correct, they demand a rewrite of human prehistory, starting from the very beginning.

Genetics is the key. DNA contains telltale traces of events in a species' past, including information about common ancestry and speciation. In theory, calculating the timing of a speciation event should be straightforward. As two species diverge from a common ancestor, their DNA becomes increasingly different, largely because of the accumulation of random mutations. The amount of genetic difference between two related species is therefore proportional to the length of time since they diverged. To estimate when the human/chimp split occurred, geneticists can simply count the differences in matching stretches of chimp and human DNA and divide this number by the rate at which mutations accumulate. This is known as the molecular clock method.

But there's a catch. To arrive at the answer you have to know how fast the mutations arise. And that leads you back to square one: you first need to know how long ago we split from chimpanzees.

To get around this catch-22, geneticists turned to orang-utans. Fossils suggest that they split from our lineage between 10 and 20 million years ago. Using this fudge, geneticists arrived at a mutation rate of about 75 mutations per genome per generation. In other words, offspring of humans and chimps each have 75 new mutations that they did not inherit from their parents.

This number rests on several big assumptions, not least that the orang-utan fossil record is a reliable witness – which most agree it is not. Even so, it led to a guess that human ancestors split from chimpanzees between 4 and 6 million years ago. When fossil-hunters hear this number, they cry foul. The lower end of the estimate is particularly hard to swallow. *Australopithecus afarensis* – an early hominin from East Africa – already has distinctly human characteristics, yet dates back at least 3.85 million years. Its canines were small and it walked upright.

Both of these traits are considered to be characteristic of hominins, not chimps. And yet it is hard to see how they could have evolved so quickly, in perhaps as little as 150,000 years after the split. In the face of the fossil evidence, a 4-million-year divergence date seemed unlikely. Even a 5- to 6-million-year-ago split was met with scepticism. Again, certain fossils from Africa date from around the same period and bear unmistakable marks of humanity. Though the interpretation of the remains is controversial, many regard them as being post-split.

Simply put, the palaeontologists were sure there was little chance that the DNA results were accurate. Humanity, they affirmed, had to be older than the geneticists claimed. History looks set to prove them right: researchers studying human populations have now been able to observe mutations almost as they happen. And that makes all the difference. Instead of relying on an estimate based on rare fossils, we can watch the molecular clock ticking in real time.

In September 2012 Augustine Kong of Decode Genetics in Reykjavik, Iceland, and colleagues published one such ground-breaking study. After scanning the genomes of 78 children and their parents to count the number of new mutations in each child's genome, they found that every child carries an average of 36 new mutations. Crucially, that is half what was previously assumed, meaning that the molecular clock ticks more slowly than we thought – pushing the human/chimp split further back in time.

How far back, exactly? That same year, Kevin Langergraber at Boston University and his colleagues solved another piece of the puzzle. Mutation rates in studies like Kong's are measured per generation. To convert this into an estimate of when our ancestors split from chimps, you need to know how long a generation is – in other words, the average age at reproduction. We have a good handle on this for humans, but not in other primates. For chimps, estimates have ranged from 15 to 25 years.

Using data from 226 offspring born in eight wild chimp populations, Langergraber found that, on average, chimps reproduce when they are 24-and-a-half years old. Based on the new numbers, his team estimated that the human lineage went its separate way at least 7 million years ago, and possibly as far back as 13 million years ago.

Soon after Kong and colleagues published their new estimate, another team – including many of the same researchers – published another. They analysed DNA from more than 85,000 Icelanders, focusing on short stretches of DNA called microsatellites – which are a more reliable record of mutations. The rate they found was not quite as slow as Kong's. As a result, their estimate of the timing of the split is a more constrained 7.5 million years.

Either way, our lineage is considerably older than we once thought.

2
On to two legs

After our ancestors split from the ancestors of chimpanzees, they carried on changing. These earliest hominins were still very ape-like, but they soon evolved a few telltale traits. In particular, they seem to have started walking differently.

Throughout the twentieth century, the hominin fossil record did not extend very far back in time. There were plenty of fossils from the last 2 or 3 million years, but older periods revealed almost nothing. For the period from the putative split with chimpanzees (which we now suspect happened between 13 and 7 million years ago) to about 4 million years ago, there was a blank. But all that changed during the first few years of the twenty-first century.

Orrorin tugenensis

In December 2000 scientists working in Kenya announced that they had discovered fossil bones of the earliest human-like creature yet found. The remains were initially dubbed 'Millennium Man' and the species was formally named *Orrorin tugenensis*. The fossils are thought to be about 6 million years old – some 1.5 million years older than previous finds.

The fossils were found in the Great Rift Valley, where many early hominid fossils have been uncovered. A local herdsman stumbled across the first specimen in October 2000, and joined the team of French and Kenyan archaeologists that excavated the site. The team went on to dig up remains from five individuals, including fragments of jaws with teeth and arm, leg and finger bones.

A beautifully preserved thighbone suggested that *O. tugenensis* had strong back legs and possibly walked upright, a key trait linking it to human-like animals, or hominids. This implied that bipedalism developed 2 million years earlier than thought. The sturdy arm bones suggested that the creature could also happily scramble around in the treetops.

But the key finding that linked *O. tugenensis* to modern humans was its teeth. It had relatively small canines and robust molars, meaning that it probably enjoyed a diet of fruit and vegetables, with occasional meat.

Later studies found stronger evidence that *O. tugenensis* was bipedal. In 2004 Robert Eckhardt of Pennsylvania State University in the US carried out a CT scan on the most complete of *Orrorin*'s three thighbones. He hoped that revealing its internal structure would indicate the biomechanical use of the bone.

Essentially, the thighbones are supporting a horizontal pelvic beam that takes the weight of the head and body. The precise load this places on the thighbones depends on body posture, and this determines the musculature and structure of the thighbones. In knuckle-walking chimps, the strong outer cortex is the same thickness at the top and bottom of the thighbone. However, bipedal, upright walking applies different forces, which means that the cortex in humans is thicker by at least a factor of four on the bottom part of the bone. Eckhardt found that the lower part of the thighbone in *Orrorin* is three times thicker than the upper – making its walking habits much closer to those of humans than chimps.

Then, in 2008, another group measured the shape of the thighbone, which reveals posture. Comparisons with thighbones of other fossils, and of modern great apes, suggest that *Orrorin* was bipedal. *Orrorin*'s particular walking style may have remained dominant for 4 million years, until the genus *Homo* evolved a stride that was better for long-distance walking and running.

The woman who found *Orrorin*: an interview

In 2000 Brigitte Senut discovered Orrorin tugenensis, *the first human ancestor known to have walked upright, in the Tugen Hills of Kenya.*

When did you know that the bones you unearthed in 2000 were important?

Immediately. The head worker at the site, Kiptalam Cheboi, found two fragments of a jaw. Other members of the team then uncovered two femurs and a humerus. We went on to study the femurs in more detail, but we could already see from their morphology that they belonged to a bipedal hominid. And from the geology of the site, we knew that it was 6 million years old. That put back the origins of bipedalism by about 3 million years, since the oldest biped known up to that point was the Ethiopian australopithecine Lucy (see Chapter 3).

The find was controversial. Why?

According to the dominant paradigm at the time, there were no hominids or human ancestors on Earth before the Pliocene, the geological epoch that began 5.5 million years ago. As soon as I laid eyes on *Orrorin*, I knew our problems were only just beginning. Sure enough, some people said we had unearthed chimpanzee remains. Everyone now acknowledges that *Orrorin* is of the human family, but there is still a debate over the relationship between *Orrorin* and Lucy. Are they both direct ancestors of modern humans, or did the australopithecines branch off at some point?

Do you consider *Orrorin* to be the notorious 'missing link'?

No; the missing-link concept implies that a great ape of the modern kind is the common ancestor, but the ancestor of *Orrorin* bore no resemblance to modern chimpanzees.

What does the name mean?

It means 'the original being' in the Tugen language of Kenya. The Tugen people have long devoted songs and dances to this mythical creature.

What did *Orrorin* look like?

It was a young adult between 1.10 and 1.37 metres tall, that is, slightly taller than Lucy. Like Lucy, it climbed trees as well as walking on two feet. But while Lucy had a small skeleton and large teeth, *Orrorin* had small teeth and a relatively large skeleton. It seems unlikely to me that a microdont such as *Orrorin* gave rise to a macrodont like Lucy, which in turn gave rise to the microdonts that were later hominids – but others disagree.

Do you know how the first individual whose remains you found died?

Some of the bones were covered in a fine layer of sodium carbonate, which made me think initially that it may have ventured on to the fragile crust of a hot soda lake – the likes of which you still find in the Rift Valley today – and then fallen through it and become trapped. However, one of the bones, a femur, also has tooth marks on it, and the top part is missing as if the leg had been ripped away from the torso at its fleshiest part. That's how a leopard tackles its prey. I think a leopard-like animal killed *Orrorin* and then carried its carcass up into a tree. From time to time, its bones dropped into the lake below. That's just one possible scenario, but it fits the facts.

Sahelanthropus tchadensis

Orrorin was a remarkable discovery. But in July 2002 another truly ancient species was announced. The wind-blown Djurab Desert of Chad had yielded a hominid skull 6 to 7 million years old. It was unearthed by Michel Brunet of the University of

Poitiers in France and his team, who had been digging in the area since 1994. The team also found lower-jaw fragments and isolated teeth from other individuals.

Named *Sahelanthropus tchadensis*, the new species was close to the common ancestor of humans and chimpanzees. It implied that the last common ancestor did not closely resemble any modern ape. Although its body and brain were the size of a modern chimp's, its face was quite different, with large brow ridges and much smaller canine teeth. From the back, the skull looked like that of a chimpanzee, whereas from the front it could pass for a 1.75-million-year-old *Australopithecus*.

The discovery also showed that early hominids had spread at least 2,500 kilometres from the East African rift valleys, which until then had yielded the best fossils. At that time *Sahelanthropus* lived, the climate in Chad was moist, leading the team to nickname the skull 'Toumai', a local word for babies born before the dry season.

Ever since the discovery, the key question has been whether Toumai walked upright. No leg bones have yet been found. However, Brunet has argued that the position where the spine entered the head was at least compatible with a bipedal posture.

The blend of features in the new fossil further challenged the old theory that hominids evolved each key trait only once in a line of descent. Instead, our evolution may have been a series of diversifications, in which anatomical features were 'mixed and matched'.

'Ardi'

When it was first discovered, *Ardipithecus* was not quite recognized for what it was. Tim White and his colleagues described it, in 1994, as *Australopithecus ramidus*: a new species, but belonging to the already well-known *Australopithecus* genus (see Chapter 3). A year later, they published a note revising their original classification and proclaiming that the fossil belonged to a new genus. It was now named *Ardipithecus ramidus*.

It took 15 years for a detailed analysis of the fossils to be published. 'Ardi' was confirmed to be 4.4 million years old, and it revealed that by that time our ancestors were upright, omnivorous and cooperative.

FIGURE 2.1 Small-brained and at home in the trees, 'Ardi' walked fully upright.

Ardi stood 117–124 centimetres tall and weighed 50 kilograms. Unlike chimps, but like the later australopithecines, Ardi walked fully upright (see Figure 2.1). This suggests that 'knuckle walking' is a recent adaptation of chimps and gorillas, not an ancient trait that our ancestors gave up. That said, Ardi seems to have also spent time in the trees. She had an opposable big toe, which would have helped her grab on to branches.

Ardi's brain was in some respects quite ape-like: it was much smaller than ours, with a volume of 300–350 cubic centimetres. However, her face jutted out much less than that of modern great apes. Her teeth give us an idea of her diet. The molars are smaller than those of the later australopithecines and perhaps suitable for cracking nuts and other hard foods. The other teeth do not look specialized. This suggests that she was an omnivore, eating a mix of ripe fruits and small animals.

A 2001 find revealed that the genus *Ardipithecus* survived for over a million years. That year, Yohannes Haile-Selassie, now at the Cleveland Museum of Natural History, Ohio, discovered bones and teeth from five individuals that lived 5.2 to 5.8 million years ago in the Horn of Africa. The remains were similar in shape and size to the 4.4-million-year-old *Ardipithecus ramidus*. But they had sharper canine teeth than later hominids, suggesting that they were a more primitive subspecies of *A. ramidus*.

They are now known as *Ardipithecus kadabba*; *kadabba* means 'basal family ancestor' in the Afar language. The shape of a toe bone suggests that *A. kadabba* walked upright.

Timeline of human evolution

From this point on, the story gets more complicated. Multiple hominin species lived at the same time, and there are arguments about which of them count as separate species. So, now seems a good time to present a timeline of human evolution to help you keep things straight.

55 million years ago (mya)
The first primitive primates
evolve, probably in Asia.

8–6 mya
Gorilla and human lineages
split and begin evolving
independently.

4 mya
Australopithecines appear. They have
brains no larger than a chimpanzee's,
but are definitely bipedal and there
is tentative evidence that they
used tools.

4.4 mya
Ardipithecus ramidus,
represented by the famous
fossil 'Ardi', lives in East Africa.

3.6 mya
Two *Australopithecus afarensis*
leave their footprints in volcanic
ash at Laetoli, Tanzania.

3.5 mya
Kenyanthropus platyops
lives in East Africa.

1.8 mya
Homo erectus evolves. It lives in
Africa, Europe and Asia – ranging
as far east as Java. It is the first hominin
confirmed to have lived outside Africa.

1.9 mya
Homo ergaster lives in
East Africa. *Australopithecus sediba*
lives in what is now South Africa.

13–7 mya
Chimp and human
lineages diverge and begin
evolving independently.

7 mya
Sahelanthropus tchadensis,
one of the first hominins,
lives in Central Africa.

5.5 mya
Ardipithecus kadabba,
a forest-dwelling hominin that
walks on two legs, lives in East Africa.

6 mya
Orrorin tugenensis,
possibly the first bipedal
hominin, lives in East Africa.

3.3 mya
The oldest known stone tools
are from Lomekwi in Kenya.

3.2 mya
Lucy, the famous
Australopithecus afarensis,
lives in what is now Ethiopia.

2.3 mya
Homo habilis appears. It may be
the first species in our genus, *Homo*.
Its brain is slightly larger than that
of earlier hominins.

2.6 mya
Oldest known
Oldowan
stone tools.

2.7 mya
Paranthropus boisei
lives in East Africa.

600,000 years ago (ya)
Homo heidelbergensis lives in
Africa and Europe. It has a similar brain
capacity to that of modern humans.

450,000 ya
The Neanderthal and Denisovan
lineages split and begin evolving independently.
The Denisovans spread throughout
Asia while Neanderthals colonize Europe.

130,000–115,000 ya
The first migration out of Africa
by *Homo sapiens* gets as far as
the Middle East.

143,000 ya
The most recent evidence
of *Homo erectus* in Asia.
They presumably die out soon after.

110,000 ya
The oldest known jewellery,
made from marine shells.

70,000 ya
The second migration out of
Africa by *Homo sapiens* goes global.

39,000 ya
The last dated evidence of
Neanderthals in Europe. After
they die out, *Homo sapiens* is
the only hominin species alive.

40,000 ya
The oldest known cave art is
a red dot in El Castillo
Cave in Spain.

18,500 ya
Modern humans are living
in the Americas by this time.

16,000 ya
The oldest accepted
evidence of domestic dogs.

400,000 ya
The oldest known wooden spears,
from Schöningen, Germany.

335,000–236,000 ya
Homo naledi lives in southern Africa.

190,000 ya
Hobbits (*Homo floresiensis*) arise
in Indonesia. They stand just over 1 metre
tall and have advanced stone
tools despite their small brains.

320,000 ya
The oldest known specimens of
Homo sapiens live in Morocco.

65,000 ya
The earliest evidence of humans
living in Australia.

50,000 ya
The proposed 'great leap forward':
culture starts to change more rapidly
than before. People begin burying their
dead, create clothes and develop
complex hunting techniques.

45,000 ya
Homo sapiens interbreeds
with Neanderthals, probably
in the Middle East.

50,000 ya
The most recent traces of
Homo floresiensis in Indonesia.
They probably die out soon after.

11,000 ya
The oldest evidence of
large-scale architecture,
in the Levant.

10,000 ya
Agriculture develops and spreads.

5,000 ya
The first true writing appears.

5,500 ya
The Bronze Age begins.
Humans begin to smelt and
work copper and tin.

3
Lucy and her sisters

The next big event in the story of our origins was the rise of the australopithecines: a group of more human-like creatures that thrived in Africa between about 4 and 2 million years ago. Australopithecus gave us perhaps the most famous single fossil in all of palaeoanthropology: the legendary 'Lucy'. These creatures were also a crucial step on the path to humanity.

The Taung Child

In 1924 Raymond Dart made a discovery that would change his life, and upend the established wisdom in anthropology.

Quarrymen had found the fossilized skull of an infant ape-like creature at a site called Taung in South Africa. Dart identified the skull as that of an early human ancestor. The fossil became known as the Taung Child, the first small-brained hominid to be discovered. Dart named it *Australopithecus africanus*, which means 'southern ape from Africa'.

At the time, most anthropologists thought that Asia was where humanity evolved. The Taung Child was the first fossil of a human ancestor found in Africa. It provided the first concrete evidence that this continent, not Asia, was the cradle of humankind. For his insight in suggesting that this little ape-like creature played a significant role in human evolution, Dart was derided and, finally, ignored. Two decades were to pass before his ideas became accepted as a central part of anthropological thinking.

Lucy

In late 1974 anthropologists were digging for fossils in the isolated Afar region of Ethiopia. One of them, Donald Johanson, spotted several bones sticking out of the ground – ground that was known to be 3.2 million years old. At the time (decades before the discoveries of *Orrorin*, *Sahelanthropus* and Ardi), this was the oldest hominin fossil ever found (see Figure 3.1).

That evening, Johanson played a Beatles recording he had brought with him, and the song 'Lucy in the Sky with Diamonds' came on. Someone in the group suggested that they could call the new fossil 'Lucy', and the name stuck. However, her scientific name is *Australopithecus afarensis*, a species distinct from *A. africanus*, the Taung Child.

What was Lucy like? Scans of her skeleton published in 2016 confirm that she had an exceptionally powerful upper body, thanks to spending a lot of time climbing trees. The finding suggests that moving in trees may have remained important

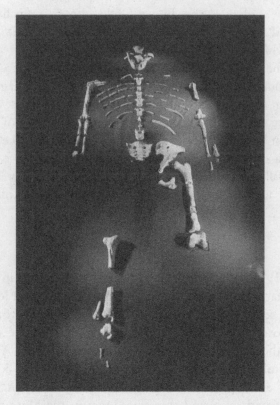

FIGURE 3.1 When the first specimen, famously named 'Lucy', was discovered in 1974, *Australopithecus afarensis* was the oldest hominin fossil ever found. Anthropologists have since identified the remains of more than 300 individuals of the species.

to some early human ancestors for millions of years after they developed the ability to walk on the ground.

Lucy had long chimp-like arms and fingers – features that would seem ideal if her life involved a great deal of tree climbing. But her legs and human-like feet show that she was what researchers call a 'terrestrial biped' – she could walk in a human-like manner. So Lucy's chimp-like arms might simply be features she inherited from a tree-climbing ancestor but no longer really used.

Later analysis found that Lucy's arm bones were thick-walled, implying that her arms were unusually strong. This is a 'use it or lose it' trait: bone strength is a direct consequence of an animal's behaviour, not something inherited. What this means is that some australopiths were equally at home in trees as on the ground – they did not have to defy their physical traits to exist in the trees. It also seems that Lucy's skeleton has injuries that suggest she died in a fall from a great height – possibly from a tall tree.

Footprints in rock

Laetoli in northern Tanzania is the site of iconic ancient footprints, capturing the moment – 3.66 million years ago – when three members of Lucy's species (*Australopithecus afarensis*) strode out across the landscape.

The Laetoli footprints were discovered in 1976. Nothing quite like them had ever been found before. They remain by far the oldest hominin footprints we know, fortuitously preserved because a group of australopiths walked across damp volcanic ash during the brief window of time before it turned from soft powder into hard rock.

In 2016 something quite unexpected came to light: the footprints of two other individuals. Researchers described 13 prints belonging to a large individual – dubbed S1 – and

---

a single print belonging to a smaller S2 australopith. S1 seems to have been walking in the same direction, at the same speed – and in all probability at the same time – as the australopiths whose footprints were uncovered in the 1970s.

It has been all too tempting to interpret the original trackways – often reconstructed as belonging to two adults and one juvenile – as evidence of a prehistoric 'nuclear family'. The new footprints show that more adults were present, including one who was much larger, the S1 individual. That has spawned a new hypothesis about australopith social groups: that they lived in societies similar to those of gorillas, with a single dominant large male accompanied by several females and offspring.

Toolmakers

A growing body of evidence suggests that the ape-like *Australopithecus* may have figured out how to make stone tools, long before the rise of modern humans.

In 2010 German researchers working in Ethiopia discovered markings on two animal bones that were about 3.4 million years old. The cut marks had clearly been made using a sharp stone, and they were at a site that was used by Lucy's species, *Australopithecus afarensis*.

The study, led by Shannon McPherron of the Max Planck Institute for Evolutionary Anthropology in Leipzig, Germany, was controversial. The bones were 800,000 years older than the oldest uncontested stone tools, and at the time few seriously thought that australopithecines had been tool users. In addition to this, McPherron had not found the tool itself.

In a 2015 study, Matthew Skinner at the University of Kent in the UK and his colleagues looked at the hands of species that would have held tools. Specifically, they looked at metacarpal bones – the five bones in the palm of the hand that connect to the digits. Because the bone ends are made of soft, spongy bone tissue, they are shaped over a lifetime of use and moulded by what that hand has done.

A chimp, for instance, spends a lot of time swinging from branches and knuckle-walking. That exerts a great deal of force on the joints in its hands, in a specific way. Skinner and his colleagues predicted how this should shape the soft bone in ape hands, then looked at modern ape bones, finding that their predictions were right.

Modern human metacarpals look different because we use our hands differently. Most of our activities involve some kind of pinching – think of how you hold a pencil or pick up a cup. This precision squeeze between thumb and fingers is uniquely human and a legacy from our flint-wielding ancestors.

When Skinner and his colleagues looked at the metacarpals of early human species and Neanderthals – who also used stone flakes for tasks like scraping and butchering – they found bone ends that were shaped like those of modern human, and unlike ape bones. Finally, they looked at metacarpals from four *A. africanus* individuals, up to 3 million years old. This revealed that their owners had been tree swingers but had also spent a lot of energy tightly pinching small objects, suggesting that they were indeed early tool users.

This is not proof that *A. africanus* used stone tools. They might have been using their strong precision grips to get at food in new ways, such as peeling tough skins off fruit. But the study does suggest that 3 million years ago – 400,000 years before the oldest known hand axes – *A. africanus* was already starting to use its hands differently from its ancestors.

You are what you eat

Australopithecines tackled a significantly different range of foods compared to their more ape-like ancestors. In fact, they might have been crackers – specialized crackers of tough nuts and seeds, that is. Australopithecines boasted mouths ideal for accessing such well-protected food, according to a 2009 study.

Australopithecus possessed jaws and teeth larger and more powerful than those of its ape ancestors. Some argued that this was for munching small, hard objects such as seeds. Others have proposed that their bigger mouths merely allowed them to eat more food with each bite. However, the 2009 results cast doubt on both explanations.

Rather than analyse microscopic cracks in tooth enamel or the chemical composition of bone, as others had done, David Strait of the University of Albany, New York and his team took an approach more common to mechanical engineering. Using a CT scanner, they measured jawbones and teeth from *A. africanus*, the species to which the Taung Child belonged. Then, with estimates of muscle strength, they calculated the maximum force that each tooth could exert before shattering.

These calculations suggest that *A. africanus*'s premolars – teeth just behind the sharper canines – were strong enough to crush the shells of nuts that would have been too large to fit between the even more powerful molars further back in the mouth. Nuts and large seeds were probably not *Australopithecus*'s favourite snack, but munching on these foods might have helped them survive lean periods.

There is also evidence that australopithecines began eating grass soon after they started leaving the trees. Australopithecines living 3 to 3.5 million years ago got more than half their nutrition from grasses, unlike their predecessors, who preferred fruit and insects. This is the earliest evidence of hominins eating savannah plants.

A 2012 study found high levels of carbon-13 in the bones of *A. bahrelghazali*, which lived on savannahs near Lake Chad in Africa. This is typical of animals that eat a lot of grasses and sedges. Previously, the oldest evidence of grass eating was from 2.8 million years ago. The 4.4-million-year-old *Ardipithecus ramidus* did not eat grass. *A. bahrelghazali* may have eaten roots and tubers rather than tough grass blades. Adding these to their diet may have helped them leave their ancestral home in East Africa for Lake Chad. The question is whether hominins moved on to savannahs permanently, or went between woodland and savannah when it suited them.

Australopithecus sediba

In 2010 the world of anthropology was rocked when another long-lost human cousin was unearthed in South Africa. Of all the australopithecine primates yet found, its anatomy is the closest to the true humans that evolved into us.

The find came in the form of two partial skeletons of *Australopithecus sediba* that were dug up in the Cradle of Humankind World Heritage Site near Johannesburg, South Africa, by Lee Berger of the University of the Witwatersrand in Johannesburg and colleagues. The skeletons are between 1.95 and 1.78 million years old.

A. sediba's physical features are closer to human than other australopithecines, but the skeletons are hundreds of thousands of years younger than the oldest fossils assigned to the genus *Homo*. To many, this implied that it was unlikely to be our direct ancestor.

Berger found the bones of a male child, aged 9 to 13 years, and an adult female in a previously unknown cave in the Malapa cave system, an area with 13 previous hominid fossil finds. Notably, the juvenile is the most complete australopithecine skeleton yet found from the period. It includes much of the skull and large parts of an arm, leg and pelvis. Both skeletons were about 1.2 metres tall and lightly built, with ape-sized brains and bodies resembling *A. africanus*, which is thought to have been a direct ancestor of humans. With long, muscular arms and strong hands, they would have been well adapted for both tree climbing and walking.

The modern human lineage is widely believed to have evolved from a line of 'gracile', or lightly built, australopithecines that goes back some 4 million years. However, the full family tree remains unclear. With the notable exception of the famed 3.2-million-year-old Lucy, an *A. afarensis*, most australopithecine and early *Homo* fossils have been very scrappy, making it hard to determine their features and relationships.

The fossils' traits do not neatly fit *A. sediba* into the hominin family tree, which includes only humans and our ancestors and extinct cousins. In 2010 Berger wrote that 'it is most likely descended from *A. africanus*', but that it was 'not possible to establish the precise phylogenetic position of *A. sediba* in relation to the various species assigned to early *Homo*'.

Timing is a key problem in determining ancestry, because the *A. sediba* fossils lived after the earliest humans. They may have been a relict population of a group that earlier gave rise to *Homo*, or a surviving sister group to the ancestral lineage.

In late 2011 Berger reported that *A. sediba* appeared to mark a halfway stage between primitive 'ape-men' and our direct ancestors. A year of detailed study had revealed that the skeletons were a hodgepodge of anatomical features: some bones looked almost human while others were chimpanzee-like. One area of particular interest was the brain size, which was small even for an australopith, with a volume of just 420 cubic centimetres.

A. afarensis, by contrast, averaged 459 cubic centimetres, despite being an earlier species. This suggests that there was no overall increase in brain size over the course of australopith evolution. But there was evidence that the brain had been subtly reorganized. The orbitofrontal region, which sits just behind the eyes, is a different shape from those of other australopiths and apes, and may have been rewired into a more human-like design.

A second group looked at *A. sediba*'s hands. The fingers are long, thin and slightly curved, just like those of apes, which would have allowed *A. sediba* to grip branches firmly. But the thumb is proportionately longer than in apes, a distinctly human trait that would have allowed *A. sediba* to grip small objects precisely. These hands may have made and used stone tools. So far, no tools have been found but, given the evidence that other australopiths used stone tools, it would not be a huge shock if *A. sediba* also did.

Other groups have focused on the pelvis, feet and ankles. They all come to the same conclusion: *A. sediba* is halfway between *Australopithecus* and *Homo*. Berger says it is not surprising that the fossil is a confusing mixture, pointing out that that is exactly what we would expect in a transitional fossil.

Eats bark, fruit and leaves

In 2012 Berger returned to the fray. His collaborators announced that they had discovered what *A. sediba* ate. On the menu: bark. It turned out that *A. sediba* had poor dental hygiene. From plaque on the fossils' teeth, the team extracted 'phytoliths' – mineral traces of *A. sediba*'s food. They found signs of fruit, bark and woody tissues.

Berger was surprised, but primatologists were not. Bark represents a considerable fraction of orang-utan diets, and

---

other primates also chew on the hard stuff: species from the golden snub-nosed monkey to chimpanzees eat bark when times are tough.

However, greater dietary surprises were in store. The team looked at sediment samples and fossilized animal faeces – coprolites – to get an idea of what the environment in which *A. sediba* lived was like. They found remnants of savannah grasses in the sediment, while pollen and woody fragments in the coprolites suggested that there might have been some woodland in the vicinity.

The team then looked at the carbon isotopes in *A. sediba* teeth to see what types of plants they ate. A 'C4' signature is typical of savannah plants like grasses and the grains they carry. These plants fix carbon in a four-carbon molecule. 'C3' indicates fruits and leaves foraged from a more forested environment.

The team expected a C4 signature – it is what most hominins have and it fits the evidence that *A. sediba* lived in an open savannah. They found the exact opposite. Clearly, the diet of *A. sediba* was different from the diet of other early hominins, but why *A. sediba* had such an unusual diet is still a mystery.

What is *A. sediba*?

The two *Australopithecus sediba* skeletons keep yielding new secrets. A set of studies published in 2013 offered further evidence that *A. sediba* may bridge the gap between the ape-like australopiths and our own genus, *Homo*.

The skeletal analyses confirmed that *A. sediba* has a mosaic of ancient australopith and modern *Homo* features. For instance, its teeth are remarkably human-like. Whereas most australopiths have large, prominent canines, *A. sediba*'s are small, like ours, according to Darryl de Ruiter at Texas A&M University in College Station and his colleagues.

The skeletons also suggest that *A. sediba* had the early makings of a tapering human waist. Peter Schmid of the University of Witwatersrand, South Africa, led a team which found that its lower ribs sweep inwards, as ours do. This allowed abdominal muscles to be arranged in a way that makes walking more efficient. Other australopiths are thought to have lacked a waist, says Schmid.

In other ways, *A. sediba* was very unlike early humans. Jeremy DeSilva at Boston University in Massachusetts and his colleagues found that its legs and feet were those you would expect of a tree climber. Humans – like most other australopiths – have a rigid foot. *A. sediba*'s foot was much more flexible, making it perfect for gripping tree trunks and branches.

This poses a puzzle. If australopiths spent more time walking on the savannah and less time in the trees than their ancestors, why was *A. sediba*, the most human-like of all australopiths, so well adapted to tree living? Conceivably, some australopiths returned to life in the trees – or it could be evidence for a deeper tree-dwelling lineage in South Africa.

4
Bigger brains

Famously, our species is known as Homo sapiens. *But we are not the only species to have borne the name* Homo – *not by a long shot. The first* Homo *species evolved more than 2.8 million years ago, long before we came along, and they are some of the most remarkable creatures ever to have lived. For one thing, they became some of the most widespread animals on Earth.*

If there is one word to describe the fossil record of the Homo *genus, it is 'confusing'. A host of species of* Homo *has been described, often on the basis of incomplete or even fragmentary skeletal remains. Vigorous arguments rage over which of them are true species and which simply represent variation within a species. It is not uncommon to find the same fossil grouped into three different species by different researchers. But if we take a step back, we can reduce the first* Homo *to three main species:* Homo habilis, H. erectus *and* H. heidelbergensis. *Between them, these three species account for a big chunk of the story.*

Beginnings

The oldest known species of *Homo* is *Homo habilis*, which was confined to Africa. The year 2015 saw the discovery of the oldest known fossil belonging to this species. Unearthed in Ethiopia, the broken jaw with greying teeth suggests that the *Homo* lineage existed up to 400,000 years earlier than previously thought. The fragment dates from around 2.8 million years ago, and is by far the most ancient specimen to bear the *Homo* signature. Previously, the earliest such fossil was one thought to be up to 2.4 million years old.

The fossil has a mixture of traits. It may pinpoint the time when humans began their transition from primitive, ape-like *Australopithecus* to the big-brained conqueror of the world. Geological evidence showed that the jaw's owner lived just after a major climate shift in the region. Forests and waterways rapidly gave way to arid savannah, leaving only the occasional crocodile-filled lake. Except for the sabre-toothed big cats that once roamed these parts, the environment ended up looking much as it does today. The pressure to adapt to this new world may have jump-started our evolution into what we see looking back at us in the mirror today.

The emerging *Homo* species probably began eating more meat and using better tools – a change reflected in a more delicate jaw unearthed in 2013. After all, if you had a nice sharp stone to cut with, there was no need for a mouth built for tearing food to shreds.

Handy man

Figure 4.1 shows a reconstruction of the skull of one of the first known members of the human genus, *Homo habilis*, which means 'handy man', from about 1.8 million years ago.

The original fossil, from Tanzania, which was first reported in 1964, is incomplete. It consists of just a few distorted fragments. But in 2015 Fred Spoor at University College London led a team to create a computer reconstruction that realigned the fragments and filled in the missing parts. This made it possible to compare the skull with other fossils from what was a critical time for early human evolution.

The digitally reconstructed skull shows that handy man shared some features with *H. erectus*, but in other ways resembled *Australopithecus afarensis* (Lucy's species), which lived some 3.2 million years ago.

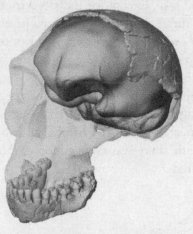

FIGURE 4.1 A reconstructed *Homo habilis* skull based on the bones of two specimens, one from modern-day Tanzania and one from Kenya

Man of grass

One of the stories that is perpetually told about human evolution is that our ancestors 'came down from the trees' and moved out 'on to open grassland'. The move down from the trees actually predates the *Homo* genus: *Australopithecus* and even the older *Ardipithecus* clearly spent plenty of time on the ground. But the move to grassland came later.

Still, by 2 million years ago humans were living and thriving on open grassland in Africa, making stone tools and using them to butcher zebra and other animals. That's according to powerful evidence from artefacts found at Kanjera South, an archaeological site in south-western Kenya.

There is no clear evidence of any hominin being associated with, or foraging in, open grassland prior to this time. Other earlier hominins that have been found in the geological record – such as *Ardipithecus ramidus* and *Australopithecus afarensis* – lived either in dense forest or in a mosaic of woodland, shrub and grasses.

The Kanjera South site offers a glimpse into the lives of our ancestors as they were starting to adapt to life on the plains. The site is a grassland setting, dominated by grass-eating animals. Tests show that the site was over 75 per cent grassland 2 million years ago. The wider area was teeming with zebras, antelope and other grazers, and all the animals carried the same telltale chemical signal suggesting that they were eating grass.

Say yes to meat

The dawn of the genus *Homo*, around 2.5 million years ago, was a watershed for another reason. This seems to have been the time when our ancestors evolved beyond their vegetarian roots and became meat eaters.

In 1999 researchers found cut marks on animal bones dated at around 2.5 million years old. But no one could be sure that they were made by meat-eating hominids, because none appeared to have suitable teeth. However, a 2013 study revealed that the first members of *Homo* had much sharper teeth than their most likely immediate ancestor, *Australopithecus afarensis*, the species that produced Lucy.

Eating meat requires teeth adapted more to cutting than to grinding. The ability to cut is determined by the slope of the cusps, or crests. Steeper crests allow a creature to consume tougher foods. The crests of teeth from early *Homo* skeletons are steeper than those of gorillas, which consume foods as tough as leaves and stems, but not meat. The crests of teeth from *A. afarensis* are not only shallower than those of early *Homo* but also shallower than those of chimpanzees, which consume mostly soft foods such as ripe fruit, and almost no meat. In other words, early *Homo* had teeth adapted to tougher food than *A. afarensis* or chimpanzees. The obvious candidate is meat.

This finding is fairly uncontroversial. But some anthropologists have advanced a much more radical idea: that early *Homo* were not just eating meat but also cooking their food. That would imply that they had discovered how to control fire.

The idea that the invention of cooking fundamentally influenced our evolutionary past was given a boost in 2003 by a study of modern diets. A team of anthropologists concluded that this newfound culinary talent was the only way to explain the huge change in our evolution 1.9 million years ago, when *Homo erectus* appeared. This was a more human-like species that was so successful that it spread out of Africa all the way to Java, Indonesia.

H. erectus was 60 per cent larger than its predecessors, and sported the largest increase in human brain size ever seen. Some experts believe that this growth spurt was fuelled by protein derived from eating raw meat. But anthropologist Richard Wrangham of Harvard University in Cambridge, Massachusetts, has long argued that it was triggered by cooking plant food, such as roots and tubers.

The heat of cooking smashes open cells and breaks down indigestible fibre into energy-giving carbohydrates. The advent of cooking would therefore account for *H. erectus* having a shorter gut and smaller teeth, and explain why early humans became more sociable as they brought food back to a central cooking area.

In support of this idea, Wrangham found that people need to eat twice as much raw food as cooked food to gain the same energy from a vegetarian diet, and 50 per cent more if their diet includes raw meat as well as plants. From a study of people in Germany who ate a raw food diet, Wrangham calculated that a person eating uncooked, vegetarian food would have to consume around 9 per cent of their body weight every day to get enough calories to maintain a leisurely modern Western lifestyle. That's more food than the average American eats on Thanksgiving Day.

The origin of cooking: an interview with Richard Wrangham

Cooking is what allowed us to become human, says anthropologist Richard Wrangham, author of Catching Fire: How Cooking Made Us Human *(2010).*

What was the central mystery of human evolution that you were trying to solve?

I was sitting next to the fire in my living room and I started asking the question, when did our ancestors last live without fire? Out of this came a paradox: it seemed to me that no human with our body form could have lived without it.

Why can't a human exist on the same diet as a chimpanzee?

A chimpanzee's diet is like eating crab apples and rose hips. Just go into the woods and find some fruits, and see if you can come back with a full stomach. The answer is you can't. The big difficulty is that the nutrient density is not very high. This is problematic for humans because we have a very small gut, about 60 per cent of the volume it would be if we were one of the other great apes. We don't have enough intestine to keep low-quality food in our gut long enough to digest it.

So cooking provided some kind of a watershed for humans to split from our chimp-like ancestors?

Yes; I believe the point at which our bodies show adaptation to cooking is 1.9 million years ago. The evidence is in the changes that took place when we evolved from ancestors that were like chimpanzees but were already standing upright. Cooking led to increased energy intake.

What was the result of having more energy?

Maximizing energy from food allowed us to lose a third of the large intestine and significantly expand our brain size. It affected our brain because humans were social and there was a premium on being as intelligent as possible in order to outsmart your opponents in competition, ultimately for mates.

The cooking row

The key stumbling block for the theory that our early ancestors cooked their food is the lack of convincing evidence that hominins could control fire more than a million years ago. This problem only got worse in 2011, when evidence emerged that humans only began controlling fire very recently.

A review of supposed archaeological hearths in Europe suggested that the oldest date was just 400,000 years ago. The finding suggests that humans expanded into cold northern climates without the warmth of fire – and that cooking was not the evolutionary trigger that boosted our brain size.

Many of the 'smoking guns' for prehistoric fire use – charred bone fragments or chunks of charcoal – do not necessarily imply that early humans could control fire. Our opportunistic ancestors may simply have exploited the occasional wildfires triggered by lightning, for example.

To try to pin down the earliest evidence of controlled fire use, researchers re-examined the data from more than a hundred European sites. They were looking for evidence of fires that were unlikely to have occurred naturally – those in caves, for example – and for clues that fire had been used in a controlled way. These include activities such as making pitch: some early hominins made this sticky substance by burning birch bark and using it to glue pieces of flint to wooden handles, to make stone tools easier to use.

The earliest European hearths date back between 300,000 and 400,000 years, the researchers concluded. Although the study investigated only European sites, evidence of controlled fire use at a number of other sites is also up for debate. The Swartkrans site in South Africa is believed by some to contain 1.6-million-year-old evidence, in the form of hundreds of charred bones. But that might just have been sporadic natural fires that people used to their advantage. In fact, just one site earlier than the 400,000-year mark has strong evidence of controlled fire use: the 780,000-year-old Gesher Benot Ya'aqov site in Israel, where charred flints, seeds and stone tools have been found.

The conclusion questioned Wrangham's hypothesis that an increase in human brain size was tied to the invention of cooking. However, the story does not end there. Another study, from 2012, presented evidence that the control of fire came at least a million years ago.

There are no obvious hearths in South Africa's Wonderwerk Cave. So, instead, researchers used microscopic analysis to study the sediments on the cave floor. They found evidence of ash and traces of burnt bone in layers that formed a million years ago. The burnt remains are 30 metres from the present entrance to the cave, so they are unlikely to represent the action of wildfires. It is more likely that hominins – probably *Homo erectus* – carried fire into the cave. The burnt bone fragments – including bits of tortoise bone – suggest, but do not prove, that *H. erectus* was cooking food.

However, this is hardly final proof of Wrangham's cooking hypothesis. The tiny traces of fire in the cave stand in contrast to the extensive ash deposits from fire found in much later sites of human occupation. This suggests that *H. erectus* was not using fire regularly, or routinely cooking food – despite its small teeth and large brain.

Homo erectus *leaves Africa*

Whatever the truth of the claim that early *Homo* species worked out how to cook food, one of these species clearly did achieve something that none of their ancestors and relatives did. From *Orrorin* through to *Australopithecus*, all the hominins were confined to Africa. There are no fossils of them from anywhere else.

But all that changed with the rise of *Homo erectus*. The first fossil of this species was not found in Africa at all, but on Java in Indonesia: hence its early nickname 'Java Man'. Evidently, some *H. erectus* managed to migrate out of Africa and spread throughout much of Europe and Asia. Yet, compared with modern humans, *H. erectus* had a small brain and could make only the simplest tools. This suggests that it did not take any great intelligence for them to go global.

Some of the best fossils of *H. erectus* outside Africa come from a site in Dmanisi, Georgia. They include an entire skull from an *H. erectus* that lived 1.8 million years ago – the earliest completely preserved specimen ever found.

An even earlier exit from Africa?

There is tentative evidence that hominins left Africa even sooner than has been thought, perhaps even before *H. erectus* evolved – but this remains a minority view.

In 2016 scientists claimed that humans were living in India 2.6 million years ago, based on an analysis of stone tools and three cow bones with cut marks. Researchers found the artefacts on the Siwalik Hills about 300 kilometres north of New Delhi, India. There, tectonic activity has exposed an outcrop of bedrock dating back at least 2.6 million years. The bones and tools were found lying on the surface. The team's examination of the cut marks on the bones suggested that they were made with a stone tool. On this basis, they claimed that hominins lived there 2.6 million years ago.

Taken at face value, the finds suggest that our *Homo* genus had migrated into Asia much earlier. Another possibility is that the earlier ape-like *Australopithecus* lived in Asia as well as Africa. But the evidence is weak. In particular, it is problematic that the stone tools and bones were found on the surface rather than in a dateable rock layer.

A link to the Neanderthals

The third crucial *Homo* species was *H. heidelbergensis*. This species lived relatively recently, probably from 700,000 to 200,000 years ago. It's generally thought that they evolved from older *H. erectus* populations, and that they gave rise to modern humans, as well as our cousins the Neanderthals and Denisovans. This species is therefore thought to have been a crucial stepping stone – the ancestor of modern humans as well as our close relatives the Neanderthals (see Chapter 5).

This story makes a certain amount of intuitive sense, because there is evidence that *H. heidelbergensis* were remarkably advanced for their time. For instance, in 2012 archaeologists found the oldest evidence of stone-tipped spears. The discovery in South Africa suggested that it was neither our species nor Neanderthals that pioneered the use of such spears, but *H. heidelbergensis*.

In line with this, there is a 400,000-year-old site near Schöningen, Germany, where wooden spears have been found associated with the remains of 19 horses. This seems to suggest that *H. heidelbergensis* mounted a carefully planned ambush there.

There is also intriguing evidence that *H. heidelbergensis* cared for invalids. In 2010 archaeologists described the most elderly ancient human ever found. He was an *H. heidelbergensis* that lived 500,000 years ago, and was about 45 years old when he died. He has been named 'Elvis', after his pelvis and lower backbone were uncovered in the Atapuerca Mountains, northern Spain. Elvis was too old to hunt and suffered terrible lower back pain. His spine was bent forward. To keep an upright posture he may even have used a cane, just as elderly people do today.

The fact that Elvis was so infirm suggests that his contemporaries must have looked after him. He could not have been physically active, but he may have had valuable knowledge that he shared with other members of the group that helped them survive. In line with this, in 2009 the same team reported evidence from Atapuerca that a 12-year-old child with skull malformations was cared for by the same group.

Ancestry in doubt

In recent years it has become possible to read DNA from preserved hominin bones, and a series of such studies has thrown the conventional *H. heidelbergensis* story into doubt.

In 2016 scientists described the oldest DNA from the nucleus of a human cell to be sequenced so far. The 430,000-year-old DNA came from mysterious early human fossils found in the Sima de los Huesos, or 'pit of bones', in the Atapuerca Mountains. The DNA revealed Neanderthals in the making. The Sima fossils look as if they come from ancestors of the Neanderthals – who evolved some 100,000 years later. But, confusingly, a 2013 study found that their mitochondrial DNA is more similar to that of Denisovans (see Chapter 5), who also lived later – thousands of kilometres away, in southern Siberia.

So who were the Sima people and how are they related to us? To find out, geneticists pieced together parts of the Sima hominin's nuclear DNA from samples taken from a tooth and a thighbone. The results suggested that they are more closely related to ancestors of Neanderthals than to those of Denisovans – meaning that the two groups must have diverged by 430,000 years ago. This is much earlier than the geneticists had expected.

It also alters our own timeline. We know that Denisovans and Neanderthals shared a common ancestor that had split from our modern human lineage. In the light of the new nuclear DNA evidence, this split might have happened as early as 765,000 years ago.

Conventional thinking is that modern humans, Neanderthals and Denisovans all evolved from *H. heidelbergensis*. However, *H. heidelbergensis* did not evolve until 700,000 years ago – potentially 65,000 years after the split between modern humans and the Neanderthals and Denisovans. Instead, another, obscure species called *Homo antecessor* might now be in the frame as our common ancestor. This species first appeared more than a million years ago.

Still, even if *H. heidelbergensis* was not our direct ancestor, it seems likely that it was quite closely related to it.

The discovery of Homo naledi

There is one more *Homo* species to discuss – the mysterious *Homo naledi*. *H. naledi* was revealed only in 2015, and research is proceeding apace – so fast, in fact, that this section of the book, more than any other, is likely to be out of date within a few years.

The story begins in the Rising Star cave system in South Africa. On 13 September 2013 two cavers, Steven Tucker and Rick Hunter, made their way down into the maze of dark passages. The pair were hoping to find tunnels that no human had ever set foot in before. Having crept up a narrow ridge known as the Dragon's Back, with 15-metre drops on either side, Hunter and Tucker arrived in a chamber thought to be a dead end. But, peering down, they discovered a narrow chute that led into another chamber.

Tucker went in first. Twelve metres down the chute, he emerged through the roof of another chamber and climbed down to the floor. The room was barely 3 metres wide. A narrow passage leading out of this chamber and on to another was just wide enough to pass through, so he called Hunter to join him. The first thing Tucker saw when they shuffled through into the next chamber was yet another passageway leading out of it. The bones came second. They were sticking out of the cave floor.

Palaeoanthropologist Lee Berger of the University of the Witwatersrand (the discoverer of *A. sediba*, discussed in Chapter 3) had been asking caving clubs to get their members to look out for fossils. So when Tucker and Hunter spotted a jawbone with what looked like human teeth, they snapped a few pictures before moving on.

Three days later, the pair were in Berger's office in Johannesburg. When he saw the pictures, Berger's jaw dropped. He immediately knew that the bones did not belong to *Homo sapiens*. Before the week was out, Berger went to see the cave. He couldn't fit down the chute, so he sent his teenage son into what is now called the Dinaledi Chamber with Hunter and Tucker. When Matthew Berger saw the bones, his hands began to shake. It was minutes before he could steady them enough to take pictures.

Just days later, on 6 October, Lee Berger posted an appeal on Facebook for palaeoanthropologists – preferably those with caving skills, and ideally with small frames. People were recruited within days – and the expedition was on.

H. naledi unveiled

Two years later, in September 2015, Berger and his colleagues published their first trove of results. The remains belonged to a previously unknown early species of our own genus, *Homo* – and they named it *Homo naledi*.

The species had a unique mix of characteristics. Look at its pelvis or shoulders and you would think it was an ape-like *Australopithecus* from 3–4 million years ago. But look at its foot and you could think it belonged to our species, which appeared within the last 500,000 years. Its skull, though, made it clear that the brain was less than half the size of ours and more like that of some species of *Homo* that lived about 2 million years ago (see Figure 4.2).

FIGURE 4.2 A replica of the skull of 'Neo', an adult *Homo naledi* with a remarkably complete skeleton

The team refers to the fossils' mixture of features as 'anatomical mosaic'. We have previously seen such a mosaic in *Australopithecus sediba*, the 2-million-year-old hominin that Berger excavated in 2008. Although it was just about possible to dismiss *A. sediba*, with its assortment of ancient and modern features, as a quirk of human evolution, the new find hints that such 'mosaicism' is not the exception in early humans but the rule.

This has implications for how we interpret other early human fossil finds representing the transition from *Australopithecus* to *Homo*. These fossils generally amount to just a few fragments rather than complete skeletons, and that might not be enough to tell us where they fit.

There is another possible conclusion to draw from the find. The sheer number of bones and their location hint at something astonishing: the bodies they belonged to appear to have been left deliberately in the cave. This has never been seen before in such a primitive human, and could have big implications for understanding the origins of modern human behaviour.

Besides a few rodent fossils and the remains of an owl that probably fell into the Dinaledi chamber by mistake, there are no other vertebrate species present. How so? Only one scenario works, the researchers argued: *H. naledi* deliberately disposed of its dead in the chamber. Perhaps the bodies were gently dropped down the shaft that researchers squeezed through to recover the bones.

There are precedents for this. At Sima de los Huesos in Spain, 28 hominin skeletons were recovered from a deep pit. But those hominins were big-brained – they looked and behaved rather like us. *H. naledi* had a brain less than half the size of ours.

The age of *naledi*

Two years later, in May 2017, Berger and his team published a slew of new results. The team had recovered 130 additional hominin bones and teeth from a second chamber in Rising Star, named the Lesedi Chamber. The additional *H. naledi* remains belong to at least three individuals, and many of the bones and teeth belong to a single, remarkably complete adult skeleton, dubbed Neo.

Perhaps more significantly, for the first time the team had worked out the age of the *H. naledi* remains in the Dinaledi Chamber: they were between 236,000 and 335,000 years old. This age range is significant. It puts *H. naledi* on the South African landscape not long before our species had begun to appear elsewhere in Africa – and long after small-brained hominins were thought to have vanished from the continent.

The age of the *H. naledi* bones also falls in a time period with a generally poor hominin fossil record. We know that several species of hominin apparently coexisted in Africa more than 2 million years ago, and that several species seem to have coexisted across Eurasia in the past 100,000 years or so. Now it seems that there was also diversity around the 250,000-year mark.

However, it is less clear where *H. naledi* fits into the human family tree. A full evolutionary analysis might conclude from those modern hands and feet that *H. naledi* branched off from other humans relatively recently. This would mean that it originated recently and then evolved to look more primitive due to its isolation.

For instance, southern Africa might have been relatively isolated from the rest of the continent, and *H. naledi*'s lineage might have had comparatively little competition from other humans. This could have relaxed the pressure to grow and maintain a large brain. If the skeleton no longer had to bear the weight of a large and heavy skull, features like the hips and shoulders might have reverted to become more like those of a small-brained hominin.

But others are reasonably sure that *H. naledi* is genuinely an early human – albeit one that survived until astonishingly recently. It could be the most primitive early *Homo* ever discovered. The species might have evolved more than 2 million years ago, as one of the earliest 'true' humans, and then survived, unchanged, for hundreds of thousands of years. *H. naledi* might be a kind of 'living fossil', a human version of the coelacanth – a primitive fish with ancestors that first appeared 400 million years ago but that is still found in oceans today. In other words, species of evolutionarily primitive humans might, in some circumstances, be able to survive for hundreds of thousands of years.

This would flip the aforementioned model on its head. Rather than seeing southern Africa as an isolated evolutionary cul-de-sac, perhaps it was actually the powerhouse of human evolution: the region where many human species (potentially including ours) first appeared.

However, this is speculation. There is one more question to ask: what ultimately happened to *H. naledi*? There are no answers to this question yet. But if the fossils really are just 300,000 to 200,000 years old, there is at least one possible scenario. Our species evolved in Africa at around that time. If those early *H. sapiens* reached southern Africa shortly afterwards, they might have contributed to the extinction of *H. naledi*.

Again, there is precedent for this. The fossil record elsewhere in the world shows that *H. sapiens* left Africa and gradually spread across Eurasia. As it did so, *H. sapiens* arrived in areas already populated by ancient humans – the Neanderthals and the Indonesian 'hobbits'. Within a few thousand years of *H. sapiens* arriving in those areas, the indigenous species of ancient humans disappeared, apparently outcompeted by *H. sapiens*.

H. naledi might have the dubious honour of being the earliest ancient human species to have been driven to extinction by the spread of our species.

Hunting hominins

Lee Berger is the palaeoanthropologist behind the recent discoveries of two new species of human ancestor. The first was *Australopithecus sediba*, and it was the sort of once-in-a-lifetime find that most people in his line of work only dream of. If Berger had taken the conventional approach, he might have built the rest of his career on analysing it.

But following convention was not what Berger, an American who has made South Africa his home, had in mind. He was convinced that even greater discoveries were waiting, particularly in the ancient caves that riddle the limestone-rich countryside. He enlisted local help to search them, and in 2013 they struck it lucky: two chambers deep inside the Rising Star cave system contained hundreds of bones from another unknown species, which his team dubbed *Homo naledi*. This time, the story generated huge publicity.

So far, his team has found the remains of at least 18 *H. naledi* skeletons, of all ages. It's a huge hoard, particularly because many hominin species exist only as a handful of bones. 'There was a real perception that these fossils are rare,' he says – and those who found them became reluctant to share access to such precious objects – 'but they are not as rare as we once thought. We were looking in the wrong places.' Asked about the scientific legacy he might leave, Berger's answer picks up on this idea. 'In 50 years, this might be looked on as the moment when we grew into an evidence-based science,' he says.

5
Our closest cousins

Our species came into existence some time within the last 500,000 years. But it was not alone. At least some of the older, more 'primitive' species like Homo naledi *were still around. What's more, several other hominin species had evolved. These were not primitives: in at least one case, the Neanderthals, they were almost as smart and capable as we are. In fact, our ancestors may not even have been conscious of a difference and, even if they were, it didn't stop them having sex with these distant relatives.*

Neanderthal minds

The most famous (and most misunderstood) extinct hominins of them all are the Neanderthals. Ever since the first fossils of a brawny, low-browed, chimp-chested hominin were unearthed in Germany in 1856, Neanderthals have stirred both fascination and disdain. German pathologist Rudolf Virchow decreed that the bones belonged to a wounded Cossack whose brow ridges reflected years of pain-driven frowns. French palaeontologist Marcellin Boule recognized the fossils as ancient, but ignored signs that the specimen he studied suffered from arthritis. It was he who reconstructed the bent-kneed, shambling brute that still lurks in the back of most people's minds. Irish geologist William King found the creature so ape-like that he considered putting it into a new genus. In the end, he merely relegated it to a separate species, *Homo neanderthalensis*.

Since then, hundreds of Neanderthal sites have been excavated. These show that Neanderthals occupied much of modern-day Eurasia, from the British Isles to Siberia, and from the Red Sea to the North Sea. Here they survived 200,000 years or more of climatic chaos before eventually disappearing perhaps as recently as 30,000 years ago.

The long-held view that Neanderthals were inferior to *Homo sapiens* is changing as, one by one, capabilities thought to be unique to us have been linked to them. What is more, the two species clearly crossed paths: publication of the Neanderthal genome in 2010 shows that they interbred. In fact, we share more than 99 per cent of our genes with Neanderthals.

If our ancestors made love, not war, the same cannot be said for the researchers who study them. The new discoveries have been pounced upon by those who believe Neanderthals thought like we did, talked like we did and enriched their world with music, decoration and symbols as we did. It has even been suggested that we are the same species. However, there are still some who vehemently argue that Neanderthal minds were no match for those of our *H. sapiens* ancestors. Surprisingly, they, too, point to the latest genetic evidence to bolster this view. So, were Neanderthals once our equal, or just another failed species of hominin?

The first pieces of evidence to support the revisionist camp come from Neanderthal lifestyles, which indicate parallels with early modern humans. We know, for example, that in addition to occupying caves and overhangs, Neanderthals also constructed shelters. Holes for wooden pegs and posts that probably supported lean-tos have been found at two sites in France. Numerous hearths dating from 60,000 years ago indicate that Neanderthals also controlled fire. They may, however, have been the first to play music around their fires. The oldest known musical instrument has been attributed to Neanderthals by its discoverer Ivan Turk, although sceptics argue that the 43,000-year-old bone 'flute' found at Divje Babe in Slovenia is just a cave bear femur punctured by wild animals.

There is also evidence that Neanderthals wore clothes. And some claim that, like today's traditional Inuit, they softened animal skins with their teeth.

Initially seen as mere scavengers, it is now clear that Neanderthals hunted formidable prey, including rhinos and fully grown mammoths. They also adapted their hunting strategies to the environment, ambushing solitary prey in forests, stalking bison and other herd animals on the steppes, and harvesting birds, rabbits and seafood at the shore.

Their toolkit, dating from between 300,000 and 30,000 years ago, required planning, concentration and great skill to make. Meticulous preparation of a stone core was needed so

that a final rap from a hammer stone would yield a prede-
termined flint flake tool. They even manufactured and used
compound tools made from more than one material, including
some of the first hafted spears, some 127,000 years ago. There
is evidence dating from 80,000 years ago that they created a
kind of glue from heated birch pitch to attach stone points to
spear hafts.

In the past it was generally believed that advances in
Neanderthal technology towards the end of their era were
simply copied from early modern humans, but research from
42,000-year-old Neanderthal sites in southern Italy refutes this.
There, at least, Neanderthals developed an array of stone and
bone tools distinct from those used by the early humans living further north. Although Neanderthals have been typecast
as incapable of change, many researchers now accept that they
did innovate.

There is also widespread acceptance that Neanderthals buried their dead. The earliest undisputed *H. sapiens* burial is in
Skhul Cave, on Mount Carmel, Israel, and dates to around
120,000 years ago. Neanderthal burials have been found at
several sites, including La Chapelle-aux-Saints in France,
where the 'Old Man' was interred with coloured earth around
60,000 years ago (see Figure 5.1), and Teshik-Tash in Uzbeki-
stan, where a nine-year-old boy was buried encircled by ibex
horns some 70,000 years ago. Dating from around the same
time are the graves of ten individuals found at Shanidar Cave
in Iraq. Ian Tattersall from the American Museum of Natural
History in New York, author of *Extinct Humans* (2001), notes
that one of these burials reveals that Neanderthals took care of
an injured individual for years before his death, providing 'powerful, presumptive evidence for empathy and caring within the
social group, and possibly for complex social roles'.

FIGURE 5.1 Neanderthals were far more advanced than was first thought. They even buried their dead, as here at La Chapelle-aux-Saints in France.

Shanidar is also the location of the famous 'flower burial'. The high concentration of pollen from medicinal plants in this grave is sometimes cited as evidence of shamanism and ritualistic funerary practices by Neanderthals. Although this interpretation has been disputed, other evidence has bolstered the case for the Neanderthals' capacity for symbolic thought.

In 2010 researchers reported that they had found perforated seashells, red and yellow pigments, and shells encrusted with a mixture of several pigments in two caves in Spain, one of them 60 kilometres from the sea. This, they claim, shows that Neanderthals adorned themselves with symbolic artefacts and that, since these date back 50,000 years, before modern humans arrived in the area, they also represent independent Neanderthal innovations.

Symbolic thought is often associated with another characteristically human trait: language. Ralph Holloway at Columbia University in New York believes that Neanderthals could speak. He has studied hundreds of brain casts from fossilized Neanderthal skulls and found that, even accounting for their big bodies, their brain size is within a few per cent of that of modern humans and, despite their sloping brows, they had frontal lobes and speech areas like ours.

As well as these physical clues, genetic tests reveal that Neanderthals had a version of a gene called *FOXP2* that is associated with language in humans. Meanwhile, fossils from Kebara Cave in Israel show that the Neanderthal hyoid, a U-shaped bone in the neck that anchors key speech muscles, matched ours.

Philip Lieberman, a linguist at Brown University in Providence, Rhode Island, agrees that Neanderthals had speech. However, he argues that before around 50,000 years ago neither Neanderthals nor modern humans could produce the full range of sounds that we can today. Having studied skulls ranging from 1.6-million-year-old *Homo erectus* through to

10,000-year-old *H. sapiens*, Lieberman concludes that neither species was capable of the vowel sounds in 'see', 'do' and 'ma'. Given this accruing evidence, many anthropologists now believe that Neanderthals probably had the same range of mental abilities as modern humans do.

You might think that the case must be closed, but some researchers still disagree with this wholesale reappraisal. Neanderthals and modern humans diverged 500,000 years ago and evolved separately in Europe and Africa. Cumulatively, that represents a million years of evolution. Given that, you would expect that there would be changes in their brains and therefore cognitive differences.

The publication in 2010 of the first Neanderthal genome lent some support to this argument. Although there is a less than 1 per cent difference between the genomes of today's humans and those of Neanderthals, this could equate to mutations in hundreds of genes.

Neanderthals may have had subtle cognitive shortcomings. It has been claimed that their lifestyles show little forward planning, and that they had less working memory capacity than modern humans, limiting the amount of information they could process at any given time. Steven Mithen, an archaeologist at the University of Reading, UK, grants Neanderthals modern capacities in knowledge of the natural world, manipulating materials and social interaction. However, he has argued that they lacked the 'cognitive fluidity' and 'capacity for metaphor' to link these domains, leaving them unable to produce complex symbolic objects.

However, for most of this period, early modern humans were not that innovative, either. There are few differences between their accomplishments and those of Neanderthals up until about 50,000 years ago. At this point, however, early modern humans pulled away, undergoing a 'big bang' of symbolic activity typified by carved statuettes, elaborate burials, an abundance of personal decorations and, eventually, elaborate cave paintings. It may be that, by the time modern humans entered Europe, they had better technology, better social organization and better brains.

Surprising talents

The list of abilities Neanderthals have demonstrated grows with every passing year, and it includes some startling tricks. For instance, they seem to be responsible for the earliest evidence of string in the archaeological record.

Perishable materials usually rot away, so the oldest string on record dates back only 30,000 years. But perforations in small stone and tooth artefacts from Neanderthal sites in France suggest that the pieces were threaded on string and worn as pendants. Similar circumstantial evidence has been found in perforated shells.

At 90,000 years old, the material purported to be string predates the arrival of *Homo sapiens* in Europe. This implies that the Neanderthals occupying the French site learned to make it themselves, rather than imitating modern humans.

Neanderthals also had the brains and guile to catch and eat birds – a skill many had assumed was beyond them. Bones found at Gorham's Cave in Gibraltar and described in 2014 suggest that Neanderthals hunted wild pigeons – rock doves – possibly by climbing steep cliffs to reach their nests. The dove bones were buried in sediments laid down between 28,000 and 67,000 years ago. Most of the excavated layers date from a time when only Neanderthals lived in the area, before the arrival of modern humans around 40,000 years ago. This means that only Neanderthals could have caught the birds.

The first artists?

Speaking of Gorham's Cave in Gibraltar, there are scratches on one part of the cave floor. They look like a hashtag, or a game of Stone Age tic-tac-toe. No one is quite sure what they really mean. But two things seem reasonably clear: the scratches are the work of a Neanderthal and they were made quite purposefully more than 40,000 years ago.

The etchings were discovered by Clive Finlayson of the Gibraltar Museum and colleagues, whose team also unearthed the rock dove bones. The age of the thick layer of clay that lies immediately on top of the rock – itself littered with Neanderthal tools and remnants of the fires they burned – tells us that the etching was made at least 39,000 years ago.

It seems clear that the etching was done with a purpose. Finlayson's colleague Francesco d'Errico from the University of Bordeaux carried out experiments to determine whether the scratches could have been made by accident. Using two kinds of Neanderthal rock points as his stylus and a slab of rock identical to the floor of the cave as his canvas, he needed to make in excess of 100 strokes to reproduce the pattern exactly.

What divides opinion is what it all means. Some say that the etchings are abstract symbols of some description, bolstering the notion that Neanderthals were capable of subtle symbolic thought. Others remain to be convinced.

Researchers have since described evidence for some rather different Neanderthal relics: a set of stone structures. In one chamber of Bruniquel Cave near Toulouse in south-west France, 336 metres from the cave entrance, there are enigmatic structures – including a ring 7 metres across – built from stalagmites snapped from the cave floor. Natural limestone growths have begun to cover parts of the structures, so by dating these growths researchers could work out an approximate age for the stalagmite constructions. They are roughly 175,000 years old, when Neanderthals were the only hominins in the region.

Today we can only guess as to why they built the circles – but the fact that they did provides a rare glimpse into their potential for social organization in a challenging environment.

The great extinction

Everyone, it seems, has a different idea about why the Neanderthals became extinct.

Those who see them as an inferior species suspect that smarter, more talkative, more social and adaptable early modern humans outcompeted Neanderthals in terms of resource use, organization and reproductive success, if not direct confrontation.

Those who believe that Neanderthals were just as smart as early humans typically look to climate change, natural catastrophes and cumulative cultural differences to explain the extinction. In his 2009 book *The Humans Who Went Extinct*, Finlayson argues that the Neanderthals relied on close-up ambush hunting, which was fine when Europe was covered in forests but became a problem when the forests shrank. As their habitat diminished, the Neanderthals became vulnerable to threats such as disease and competition.

Of course, they may also just have been unlucky.

Perhaps the first issue to settle is when the Neanderthals died out, as that should help us eliminate a few possibilities. A 2014 reassessment of major archaeological sites suggests that instead of dying out 23,000 years ago, as many had believed, Neanderthals were gone as early as 39,000 years ago. It also looks as if we shared their territory for 5,000 years, steadily replacing them as we spread across Europe.

This seems to support the idea that our direct ancestors pushed Neanderthals out.

The researchers who carried out the reassessment used improved techniques to date material from 40 key sites in Europe, spanning the period when humans reached Europe and Neanderthals vanished. Every possible or definite Neanderthal site was at least 40,000 years old. In other words, Neanderthals had largely, and perhaps entirely, vanished from their known range by 39,000 years ago. There are Neanderthal artefacts claimed to be 23,000 years old, but the team could not get any solid dates from them, so while a late survival is possible, there is no real evidence.

But that does not mean we murdered our cousins. There is no evidence humans ever killed Neanderthals, and they may not have met very often. So what role did we play? Many now suspect that we were the last straw for an already fragile species. For much of their 400,000-year history, Neanderthals were few and far between, according to a 2009 analysis of their genetic material.

That conclusion isn't earth shattering. Archaeological digs suggest that Neanderthals hardly lived in megacities. It's difficult to put a number on the population of a species based on DNA alone, but less than a few hundred thousand of these archaic humans roamed Europe and Asia at any one time.

What is most obvious is how little genetic heterogeneity they possessed. The mitochondrial genomes of six Neanderthals recovered in Spain, Croatia, Germany and Russia differ at only 55 locations out of more than 16,000 letters. This represents one-third of the mitochondrial diversity of modern humans. Because of this low diversity, Neanderthal populations must have been relatively small.

The researchers analysed bone samples that, by and large, came during the twilight of the Neanderthal's reign around 40,000 years ago. They may have obtained a genetic snapshot of a species on the verge of extinction. However, other genetic clues indicate that Neanderthal populations stayed low for much of their history.

The power of wolves

US palaeoanthropologist Pat Shipman believes that one key to our success at displacing the Neanderthals was our partnership with a weapon that wagged its tail: the domestic dog. Thanks to dogs, we may have been better at hunting than they were.

Until recently, no one thought domestic dogs appeared until about 15,000 years ago. But in 2009 a research team began investigating ways to tell dogs apart from wolves using statistical methods. These two canids are so similar that they can and do interbreed; no simple genetic or physical trait distinguishes them. However, a complex analysis of skull shape reliably separates wolves from both modern dogs and from accepted prehistoric wild dogs. Analysing additional fossil canid skulls, the team recognized a group of ancient dog-like animals intermediate in shape between wolves and prehistoric dogs. Shipman calls them 'wolf-dogs', not because she believes they were hybrids but because deciding which group they belonged to is not easy.

Whatever wolf-dogs were, they were different from contemporary wolves. Chemical analysis of their bones shows that their diets differed from those of humans or wolves at the same sites. Wolf-dog mitochondrial DNA differs from that of any other canid and is very primitive compared with that of other modern and fossil dogs and wolves.

The oldest wolf-dog yet identified is 36,000 years old, much older than expected for a domesticated animal. All known wolf-dogs occur in sites created by humans, not Neanderthals. The sites contain the bones of dozens, even hundreds, of woolly mammoths, although mammoths were previously rare in archaeological sites. Some were clearly hunted, their bones butchered, skinned and charred. The sites include hearths, tools and huts built from mammoth bones.

Although top predators are always rare in ecosystems, wolf remains at these sites are so abundant that they must also have been targeted. Their luxuriant fur would have been useful in near-Arctic conditions, and territorial wolf-dogs – like wolves and dogs today – would probably not tolerate the presence of any other canid.

Even if wolf-dogs were poorly domesticated, cooperating with them would have offered huge advantages during a hunt.

Written in the genes

In 2010 the story of human evolution had to be rewritten. Geneticist Svante Pääbo of the Max Planck Institute for Evolutionary Anthropology in Leipzig, Germany, and his colleagues announced that they had sequenced the genome of a Neanderthal. That is, they had managed to read all the genes of a Neanderthal – even though it had been dead for tens of thousands of years. This was an astonishing feat. Pääbo's team had pioneered the genetic study of Neanderthals but an entire genome had never been obtained.

It turned out that every human whose ancestral group developed outside Africa has a little Neanderthal in them – between 1 and 4 per cent of their genome. In other words, humans and Neanderthals had sex and had hybrid offspring. A small amount of that genetic mingling survives in 'non-Africans' today: since Neanderthals did not live in Africa, sub-Saharan African populations show no trace of Neanderthal DNA.

The interbreeding seemed to have happened around 50,000 years ago. That is because all non-Africans – be they from France, China or Papua New Guinea – share the same amount of Neanderthal DNA, suggesting that interbreeding occurred before those populations split. The timing makes the Middle East the likeliest location: humans who were leaving Africa could have met resident Neanderthals and done the deed.

Even more Neanderthal sex

Since 2010 we have learned that there have been several occasions when humans and Neanderthals interbred: not just that one period 50,000 years ago in the Middle East.

For instance, it turns out that some Neanderthals carried our DNA. One group had a big chunk of modern human DNA right in the middle of the gene that may have a role in language development, called *FOXP2*. What's more, they got that DNA from us at least 100,000 years ago. This might have happened in the Arabian Peninsula or the eastern Mediterranean, based on tentative archaeological evidence that modern humans were living in those regions by then.

In 2017 evidence emerged of an even earlier inter-breeding episode. DNA from modern humans turned up in a Neanderthal fossil in Germany from 124,000 years ago. This was odd, because modern humans were not supposed to have reached Europe until about 60,000 years ago. The proposed explanation is that there was a previous migration of early humans – more than 219,000 years ago.

This would mean that modern human ancestors must have interbred with Neanderthals before 219,000 years ago – and must have migrated out of Africa into Europe much earlier than we thought.

The Denisovans

The ability to reconstruct DNA from extinct organisms has not just enabled us to study known species. It has also revealed an entire new hominin species.

The story begins in 2008, with a tiny fragment of finger bone discovered in Denisova Cave in the Altai Mountains of southern Siberia. Michael Shunkov from the Russian Academy of Science bagged and labelled the shard, and sent it off for analysis. At his lab in Leipzig, Svante Pääbo was perfectly placed to show that it belonged to a Neanderthal.

But they were all in for a surprise. The Siberian genome was quite unlike the Neanderthal's and it did not match that of any modern human. It was something completely new. Here was evidence that a previously unimagined species of humans had existed some 50,000 to 30,000 years ago – around the time when our own ancestors were painting masterpieces in the Chauvet Cave in France.

A few years on, the new species has a moniker – Denisovan, after the cave. Our picture of these mysterious people is still being painstakingly pieced together. That first sliver of bone, together with a few teeth, is all we have to go on – there is still no body – but what these meagre remains have revealed is remarkable.

Within months, David Reich of Harvard Medical School in Boston, Massachusetts, working with Pääbo, had a draft of the Denisovan genome. It showed that the Denisovans were a sister group to Neanderthals. Their common ancestor branched off from our lineage, perhaps around 600,000 years ago. Then the Denisovans split from Neanderthals some 200,000 years later, perhaps parting ways in the Middle East, with Neanderthals heading into Europe and Denisovans into Asia. Given how recent the Denisova Cave specimen is, it is quite plausible that the Denisovans were around for some 400,000 years – much longer than modern humans have existed so far.

With the bone sliver proving so enlightening, the hunt was on for more remains. In 2010 DNA analysis of a forgotten tooth found in the Denisova Cave in 2000 revealed that it, too, was Denisovan. Suddenly, there were two fossils.

Archaeologists love teeth because they can reveal so much about an animal's body and habits, especially its diet. The specimen, a third molar – a wisdom tooth from the back of the mouth – should have been a vital clue, but it was singularly baffling. At almost 1.5 centimetres across, it is huge. That marks it as primitive: our apelike ancestors had larger teeth because they needed to grind up tough food like grasses. But by 50,000 years ago humans were eating softer foods and their teeth had shrunk. The Denisovan tooth looks like a throwback. Still, hominins with unusual teeth do sometimes crop up, and wisdom teeth are the most variable in the jaw, so this enormous one could simply have been an anomaly.

Then, in August 2010, archaeologists found another large tooth. Bence Viola of the University of Toronto, who was present, thought it belonged to a bear, but genetic analysis showed it to be Denisovan. It, too, was a wisdom tooth, although from a different individual, strengthening the case that Denisovans had weirdly big teeth. That hints at a fibrous, plant-based diet, but evidence for this idea is still lacking.

In 2017 a fourth tooth was identified as Denisovan: a worn milk tooth lost by a girl of 10 to 12 years old. Unearthed in the Denisova Cave in 1984, the tooth came from a geological layer formed between 227,000 and 128,000 years ago, making it potentially the oldest of the specimens.

Meanwhile, the genome had already revealed another secret about the Denisovans – that some of us carry their genes. To find out whether humans and Denisovans interbred, the geneticists looked at the few parts of the genome that vary from person to person, searching for individuals who carry Denisovan versions of these sections. Most of the people they sampled had no sign of Denisovan DNA, even if they were from mainland Asia, where our ancestors might have been expected to run into Denisovans. However, the researchers also sequenced the genome of someone from Papua New Guinea – and found a huge signal. Other Melanesian people also carried Denisovan DNA, with an average 4.8 per cent of their genome coming from Denisovans.

Clearly, interbreeding did occur. But if Denisovans lived in southern Siberia, how on earth did their DNA end up in Melanesia, thousands of kilometres away across open sea? The most obvious explanation is also the most startling: Denisovans ranged over a vast swathe of mainland Asia and they also crossed the sea to Indonesia or the Philippines. This means that they had a bigger range than the Neanderthals.

Alternatively, perhaps they interbred with modern humans on mainland Asia, and the descendants of such encounters later moved south-east, leaving no trace on the mainland. This would mean that the Denisovans were not as widespread as all that.

To find out which theory was correct, Reich sequenced the genomes of indigenous peoples from Asia, Indonesia, the Philippines, Polynesia, Australia and Papua New Guinea. If the interbreeding had happened on mainland Asia before people populated the islands, then people on all those islands should carry some Denisovan genes. But if the Denisovans had reached the islands and interbred with humans already there, some isolated populations might be Denisovan-free. He found the latter pattern, so it is unlikely that interbreeding happened on the mainland.

The genetics, then, is telling us that the Denisovans mated with early modern humans somewhere in South East Asia. If that is true, these people were formidable colonizers. From their origins at the split with Neanderthals, they appear to have made it out of the Middle East, spreading both north into Siberia and east to Indonesia and on to Melanesia.

During the last ice age, between 110,000 and 12,000 years ago, South-East Asia would have been an especially good place to live. Instead of lush forests, there were open grassy spaces. The ice at the poles locked up lots of water, lowering sea levels by tens of metres. As a result, Sumatra and Borneo were part of the mainland.

In other words, we have had the story of the Denisovans backwards: they may be named for a cave in Siberia, but that was not their usual abode. Instead, South-East Asia was their centre (see Figure 5.2). When conditions were good they expanded north, and when conditions were bad those populations died out or disappeared.

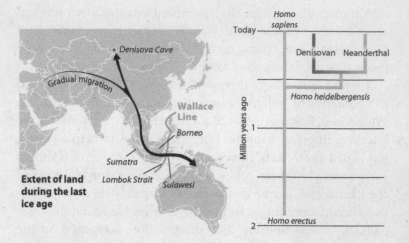

FIGURE 5.2 Denisovan fossils first turned up in Siberia, but these ancient humans probably arose in the Middle East before migrating north and south-east. Today their DNA is found mostly in people east of the Wallace Line.

That first finger-bone fragment has divulged a wealth of genetic information, but there are key questions it cannot address. For instance, were Denisovans relatively simple-minded like their *H. heidelbergensis* ancestors, or did they have the higher mental abilities of Neanderthals and early modern humans? DNA analysis cannot answer that, because we do not understand the genetic changes that made modern humans. But a skull with a big or small braincase would tell us. So the biggest challenge remains the same: to find a body.

The Neanderthal and Denisovan within

As soon as it became clear that so many modern humans carry Neanderthal and Denisovan DNA, the immediate question was: so what? Did the non-human DNA actually do anything, or was it effectively neutral – perhaps so similar to our own that its presence made no difference?

Recent genetic decoding suggests that it partly accounts for differences in our physical appearance and affects our health. Some of this 'undead' DNA even helped us survive in places for which we were otherwise ill equipped. Here are four examples:

1 **High-altitude survival**

Perhaps the most dramatic evidence of modern benefits from ancient DNA is seen in Tibetans, about 80 per cent of whom carry a particular piece of Denisovan DNA. The stretch of ancient DNA overlaps with the *EPAS1* gene, the Denisovan version of which seems to help people survive the low-oxygen conditions at high elevations. It is possible that Denisovans were adapted to life at altitude and passed the trait to humans.

2 **The gift of immunity**

The first *Homo sapiens* to arrive in Europe and Asia would have encountered new parasites and pathogens, leaving them at risk from deadly diseases. Bumping into Neanderthals and Denisovans may have had very real advantages for these early explorers: modern Eurasians owe a significant portion of their immune system genes to these two ancient relatives.

Several studies have compared these genes in living humans with the versions in our extinct cousins and found remarkable similarities. In some populations in Papua New Guinea, for example, Neanderthal and Denisovan versions of the *HLA-A* immune system gene are nearly ubiquitous. Acquiring these genetic variants may have facilitated the health and survival of migrating bands of modern humans, when they were first exploring outside Africa.

3 **Paler skin for northern skies**

Neanderthal DNA may have contributed to the appearance of pale skins in some populations. Seventy per cent of people with European ancestry carry a segment of Neanderthal DNA on chromosome 9. This DNA spans a gene associated with pale skin pigmentation and freckling, and is absent in those with no European origins.

Of course, other factors could have contributed to the evolution of paler European skin colour. But if the Neanderthal DNA did play a role, it is easy to imagine how the trait might have been beneficial to early Europeans. Dark skin protects us from the harmful effects of UV radiation, but makes it harder to generate vitamin D at higher latitudes. Pale skin could have helped our species survive on less sunlight as it migrated north.

A similar genetic legacy remains in Asia: a chunk of Neanderthal DNA on chromosome 3 is found in about half the population. It overlaps with *HYAL2*, a gene involved in skin repair. Some studies show that *HYAL2* responds to UV light exposure, which could suggest that the Neanderthal DNA helps repair skin damage from sunlight.

4 **A tolerance of cold**

One particular piece of Denisovan DNA found on chromosome 1 is present in nearly all indigenous Inuit Greenlanders. According to one study, the DNA contains two genes – *TBX15* and *WARS2* – and alters the way they are expressed in the body. *TBX15* helps generate brown fat cells that produce heat in cold temperatures, and it is thought that both genes may influence body-fat distribution.

It is possible that modern humans acquired this DNA from a lineage that had adapted to cold conditions. When they began moving into the polar regions, individuals with the ancient DNA would have had a much greater chance of surviving and passing the genes on to their children.

The man rewriting human evolution: an interview with Svante Pääbo

Svante Pääbo is a Swedish biologist famous for his work on ancient genomes. One of the founders of paleogenetics, he has worked extensively on the Neanderthal genome. His DNA analysis of a finger bone found in the Denisova Cave in Siberia suggests that the bone belonged to a previously unrecognized member of the genus Homo: *the Denisovans.*

FIGURE 5.3 Svante Pääbo

Tell us about the discovery of the Denisovans.

We knew people had lived in this cave, but thought they were either Neanderthals or modern humans. When we sequenced the DNA I was in the US, so a postdoc called me to tell me the results. He said: 'Are you sitting down?' because it was immediately clear that this was some other form of human – not a Neanderthal, not a modern human. We were totally shocked.

Does the discovery of the Denisovans raise the possibility that we once shared the planet with other types of extinct human?

Yes; I wouldn't say that's impossible but I would still guess that there would be a limited number. At the time when modern humans came out of Africa, say 50,000 years ago, what was around? Well, we know there were Neanderthals and Denisovans, and we know there were 'hobbits', or *Homo floresiensis*, the short hominins discovered on the Indonesian island of Flores in 2003. So there were at least three forms. Maybe there were a couple more; it's possible. We will know when we have studied more sites.

Do you ever wonder what the world would be like if Neanderthals had survived?

I think it's a fascinating thing to think that, with just 2,000 more generations, Neanderthals would still be here with us. Would they live in suburbia or would they live in a zoo? How would we deal with them? Perhaps racism against Neanderthals would have been even worse than the racism we experience today, because they were truly different from us in some respects. Or would having another form of human around have allowed us to be more open-minded and not make this enormous distinction we make between humans and animals today? No one can know, but it's interesting to speculate.

Meet the hobbits

There is one more *Homo* species to discuss. In 2003 the remains of a tiny and hitherto unknown species were discovered in Indonesia. The discovery was heralded as the most important for 50 years, and has radically altered the accepted picture of human evolution. Because the species was so small, they were quickly nicknamed 'hobbits' after the diminutive people in J. R. R. Tolkien's novels *The Hobbit* and *The Lord of the Rings*.

The skull and bones of one adult female and fragments from up to six other specimens were found in the Liang Bua limestone caves on Flores Island, which lies at the eastern tip of Java. The female skeleton, known as LB1 – or by the nickname 'Ebu' – was assigned to a new species: *Homo floresiensis*. The skeleton shows a mixture of characteristics that until now have been associated with very different stages of human evolution.

One of the most striking characteristics of LB1 is her height. At around a metre tall, she is far shorter even than modern Pygmies, who range from 1.3 to 1.4 metres, and roughly the same size as the relatively primitive *Australopithecus*. But australopithecines, such as Lucy, lived in Africa between 1.4 and 4.5 million years ago, whereas LB1 lived between 74,000 and 95,000 years ago.

Indeed, the shape of LB1's skull is more like that of our ancestor *Homo erectus*, which lived between 1.8 million and

200,000 years ago. That suggests that she, like *H. sapiens*, is a direct descendent of *H. erectus*. For instance, she has the protruding brow ridges typical of *H. erectus*.

Meanwhile, the body shape of *H. floresiensis* is in many ways more like the australopithecines than any human species. Her arms are so long that her hands reach almost to her knees. She has short legs with curved thighbones and a small pelvis.

But it's the size of the skull that most shocked anthropologists. It could not have contained a brain any bigger than a grapefruit – similar in size to a small chimpanzee's brain – which raises the conundrum of how such a minuscule brain was capable of the sophisticated behaviour suggested by the discovery of tools and animal remains nearby. The cut-off for the genus *Homo* used to be 500 cubic centimetres; yet *H. floresiensis* has a capacity of only 380 cubic centimetres.

A handful of stone tools from the same period were also found in the caves, along with the bones and teeth of several dwarf stegodons, an ancestor of the modern elephant. Other animal remains, including rats, bats and fish, show signs that they were cooked at around the time *H. floresiensis* inhabited the caves.

The natural question to ask is: when did the hobbits live, and when did they die out? When the species was first described, accelerator mass spectrometry dating suggested that LB1's remains were 18,000 years old, and the scientists believed that some bone fragments could be as young as 13,000 years old. The oldest remains from the site were 78,000 and 94,000 years old.

Those dates would mean that *H. floresiensis* survived well beyond the last Neanderthals, and probably after the Denisovans, too. What's more, *Homo sapiens* are thought to have colonized Flores island between 55,000 and 35,000 years ago, implying that the hobbits coexisted with us on the island for thousands of years.

That conclusion held for over a decade. But in 2016 it was demolished by a new dating study. This revealed that the skeletal remains were between 100,000 and 60,000 years old, and that the most recent stone tools were 50,000 years old. This implies that the hobbits disappeared around 50,000 years ago. That is a suspect date, because it is also around the time when modern humans arrived on Flores. It implies, but does not prove, that we may have – even if entirely inadvertently – pushed them to extinction.

What exactly is a hobbit?

The discovery of *Homo floresiensis* was so extraordinary – particularly coupled with the initial claims, since disproved, that it survived until 12,000 years ago – that some anthropologists did not believe it was a new species at all. Instead, they argued that the remains were those of a *Homo sapiens* with a disorder such as microcephaly.

The result was an almighty row that lasted for years. On one side was Robert Martin of the Field Museum of Natural History in Chicago, who argued that the existence of a species of small-brained dwarf human was a fantasy. Instead, he argued, the fossil is merely a Stone Age human with a mild form of microcephaly, a disease that stunts brain development and is associated with small stature. And he argued that the stone tools found at the site were made by regular *Homo sapiens*.

For many researchers, a 2005 analysis of the diminutive cranium confirmed that *H. floresiensis* was a unique species. It revealed remarkably advanced features for such a small brain. A 2007 study came to the same conclusion. It stated that hobbits had wrist bones almost identical to those found in early hominids and modern chimpanzees, and so must have diverged from the human lineage well before the origin of modern humans and Neanderthals.

Not everyone is convinced, but for most anthropologists *H. floresiensis* is genuine.

The origins of hobbits

Instead of *Homo* following a simple evolutionary path culminating in modern humans – *Homo sapiens* – the discovery of the hobbits suggested that early humans branched into many more forms than previously thought. This has been confirmed by the subsequent discoveries of the Denisovans and *H. naledi*.

But who were the hobbits' ancestors? Two studies from 2009 outlined the possible answers. One idea is that hobbits evolved from a very early *Homo* species, something small like *H. habilis*. Alternatively, a group of later, larger *H. erectus* could have reached Flores about a million years ago, only to shrink because of peculiar conditions on the island.

Since island species, separated from their mainland kin, often decrease in size over evolutionary time, the 'hobbit as separate species' arguments have focused on its small brain. Lack of resources could select for smaller and smaller bodies, and some researchers have argued that power-hungry brains can shrink even more drastically than the rest of the body. Put simply, it might be advantageous for the animal not to have to maintain such a big brain. On that basis, some have argued that *H. floresiensis* was an insular dwarf of *H. erectus*.

However, others say that *H. floresiensis* cannot be a shrunken *H. erectus*. For a start, the feet of *H. floresiensis* are far longer than would be expected of 1-metre-tall *H. erectus* or *H. sapiens*. Instead, perhaps the hobbit's closest relative is a species of human more ancient than *H. erectus*, with a smaller brain – perhaps *H. habilis*.

The story took a new twist in 2016: a new cache of hobbit-like remains had been uncovered on the island of Flores – six teeth, a fragment of jawbone and a tiny piece of skull. Their discoverers argued that they backed the shrunken *H. erectus* theory.

The fossils were collected in the So'a Basin on Flores, which was an African-like savannah at the time. They are 700,000 years old, much older than the original hobbit remains, suggesting that they are the ancestors of hobbits. The similarities with the hobbit are striking. In particular, the jawbone, which the team says belonged to an adult, is just as small as its hobbit equivalents. If the fossils are, in fact, older members of the hobbit lineage, then Flores seems to have been their home for hundreds of thousands of years.

Yet another idea came to the fore in 2017. The most comprehensive analysis yet suggested that the hobbits were, in fact, descended from a mystery ancestor that lived in Africa more than 2 million years ago. Some members of this ancestral group remained in Africa and evolved into *H. habilis*. The others moved out of Africa about 2 million years ago – before *H. erectus* did – and arrived in Flores at least 700,000 years ago.

Colin Groves at the Australian National University found that *H. floresiensis* was far more closely related to *H. habilis* than to *H. erectus* or *H. sapiens*, suggesting that it came from an ancient lineage and shared a common ancestor with *H. habilis*. Its more primitive, diminutive body type reinforces this idea. Groves argued that the hobbit's ancestors probably died out across Asia when bigger, more complex human species like *H. erectus* and *H. sapiens* later emerged from Africa. *H. floresiensis* was probably able to cling on in Flores for as long as it did only because of its isolation. There is no fossil evidence to indicate that *H. erectus* ever made it to the island.

This debate looks set to run and run.

More species to come

It seems unlikely that we have discovered every hominin species that ever existed. In particular, evolutionary biologists have long predicted that new human species will start appearing in Asia as we begin to look at fossilized bones found there.

A distinctive skull was unearthed in 1979 in Longlin Cave, Guangxi Province, China, but it was only fully analysed for a 2012 study. It has thick bones, prominent brow ridges and a short, flat face, and it lacks a typically human chin. To sum up, it is anatomically unique. Darren Curnoe at the University of New South Wales in Sydney, Australia, who studied the skull, has argued that it presents an unusual mosaic of primitive features, like those seen in our ancestors hundreds of thousands of years ago, with some modern traits similar to those of living people.

Then Curnoe and Ji Xueping of Yunnan University, China, found more evidence of the new hominin at a second cave – Maludong in Yunnan Province. Curnoe has dubbed the new group the Red Deer Cave people because of their penchant for venison.

Exactly where the Red Deer Cave people belong in our family tree is unclear. They could be related to some of the earliest members of our species, *Homo sapiens*. However, they could also represent a new evolutionary line that evolved in East Asia in parallel with our species, just as Neanderthals did. It is also conceivable that they are the product of matings between modern humans and Denisovans. Although we do not know exactly where they came from, we do know that the Red Deer Cave people survived until relatively recently. Some of the fossils are just 11,500 years old.

In 2015 Curnoe published an analysis of a hominin femur, also found in Maludong. It shows evidence of having been burned in a fire that was used for cooking other meat, and has marks consistent with it being butchered for consumption. It has also been broken in a way that is often used to access the bone marrow. Someone seems to have cooked and eaten it.

Things got interesting when the team tried to identify the femur. The sediment in which the bone was found dated to just 14,000 years ago, but the femur resembles the earliest members of *Homo*. This suggests that remarkably primitive-looking humans shared the landscape with very modern-looking people – even at a time when China was developing early farming.

6

Global conquerors

Today, humans are found on every continent on Earth. But it was not always thus. Although not everyone agrees, our species seems to have had its origins in Africa, and to have spread from there. But even the date of our species' origin is tricky to pin down, and the tale of how we went global is more puzzling still.

The origin of our species

For much of the twentieth century, *Homo sapiens* was thought to have evolved just 100,000 years ago. However, since the 1990s a consensus has grown that anatomically modern humans emerged in Africa at least 200,000 years ago.

This shift in thinking began in 1987. Using genetic analysis to construct an evolutionary tree of mitochondrial DNA – genetic material we inherit solely from our mothers – researchers found that we can all trace our ancestry back to a single woman who lived in East Africa some 200,000 to 150,000 years ago – the so-called 'mitochondrial Eve'.

The case for such early origins has since been boosted by fossil evidence. In 2003 researchers dated fossil remains of *H. sapiens* from Herto in Ethiopia at about 160,000 years old. Two years later another team pushed our origins even further back, dating fossil remains found at Omo Kibish, Ethiopia, to 195,000 years ago.

But some researchers have long suspected that the roots of our species are even deeper, given that *H. sapiens*-like fossils in South Africa have been tentatively dated to 260,000 years ago. Fossils found in Morocco in 2017 suggest that the *H. sapiens* lineage became distinct as early as 350,000 years ago – adding as much as 150,000 years to our species' history.

This new evidence came from a site called Jebel Irhoud, where hominin remains had been found in the 1960s, although nobody then could make sense of them. So scientists returned to Jebel Irhoud to try to solve the puzzle. In fresh excavations, they found stone tools and more fragmentary hominin remains, including pieces from an adult skull. An analysis of the new fossils, and of those found at the site in the 1960s, confirmed that the hominins had a primitive, elongated braincase. But the new

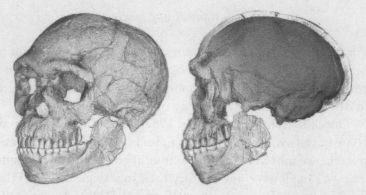

FIGURE 6.1 A reconstruction of the earliest known *Homo sapiens* fossils

adult skull showed that the hominins combined this ancient feature with a small, lightly built 'modern' face – one virtually indistinguishable from that of *H. sapiens* (see Figure 6.1).

Another study examined the stone tools. Many of them had been baked – probably because they were discarded after use and then heated when the hominins set fires on the ground nearby.

This heating 'resets' the tools' response to natural radiation in the environment. By assessing the levels of radiation at the site and measuring the radiation response in the tools, researchers established that the tools were heated between 280,000 and 350,000 years ago. They also redated one of the hominin fossils found in the 1960s and concluded that it is 250,000 to 320,000 years old.

Armed with these dates, the Moroccan hominins seem easier to understand. The researchers suggested that *H. sapiens* had begun to emerge – literally face-first – between about 250,000 and 350,000 years ago.

Going global

Long before the Nike logo and McDonald's golden arches strad-
dled the planet, our own species had penetrated every corner of
the globe. But it took a while. It seems that our species remained
in Africa for tens of thousands of years before going global: 'all
dressed up and going nowhere', as one archaeologist put it.

Skeletal remains from Skhul and Qafzeh in Israel dating from
120,000 to 90,000 years ago are the oldest confirmed traces of
modern humans outside Africa. Discovered in the 1930s, these
were once thought to represent the leading edge of a successful
wave of colonization that would take our newly evolved species north and west into Europe and, eventually, eastward across
the globe.

However, all evidence of human habitation beyond Africa
disappears around 90,000 years ago, only to emerge again much
later. The finds in Israel are widely believed to represent a precocious but short-lived surge of humanity into the wider world.
Yet the crucial questions remain: why did humans leave Africa
when they did, and what enabled them to achieve world domi-
nation this time, where previous migrations had petered out?

Richard Klein, an anthropologist at Stanford University in
California, has championed the idea that fully modern behaviour appeared in a relatively sudden burst around 50,000 to
40,000 years ago. Such behaviours encompass the manufacture
and use of complex bone and stone tools, efficient and inten-
sive exploitation of local food resources and, perhaps most significantly, symbolic ornamentation and artistic expression. This
'cultural great leap forward' tipped humans over into modernity and equipped them with the creativity, the skills and the
tools they needed to conquer the rest of the world.

By contrast, other researchers believe that the behavioural modernity that underpins the human success story evolved much earlier. They point to a growing array of artefacts that date back to 80,000 years ago (see Chapter 8). It is possible, however, that such finds might simply reflect a gradual accumulation of more modern behavioural patterns rather than the appearance of fully modern minds. If you look at the archaeological record between 100,000 and 40,000 years ago, you find occasional artefacts that seem to be indicative of modern behaviour, but they remain rare.

In 2006 Paul Mellars of the University of Cambridge proposed a new model to explain the out-of-Africa diaspora that aimed to tie together these controversial archaeological remains with recent genetic findings. The key pieces of the puzzle were genetic studies that pointed to a series of population explosions, first in Africa and later in Asia and then Europe.

Rapid population growth leaves a telltale signature in the number of differences in mitochondrial DNA between pairs of individuals within a specific population: as the time since the population explosion increases, so do the DNA mismatches. This analysis shows that African populations were increasing rapidly 80,000 to 60,000 years ago, neatly matching the evidence for an early flowering of behavioural modernity.

According to Mellars's model, human behaviour was altering between 80,000 and 70,000 years ago in ways that led to major technological and social changes in southern and eastern Africa. Key innovations, including improved weaponry for hunting, new use of starchy wild plants for eating, the expansion of trading networks, and possibly the discovery of how to catch fish, enabled modern humans to make a better living off the land and sea. Mellars argued that all this led to a massive and rapid population expansion, perhaps in just a small source region in Africa, between 70,000 and 60,000 years ago. This growing population, equipped with more complex technology, was finally able to push out of Africa and into southern Asia from around 65,000 years ago.

The human story has always been hotly contested. Now, at last, a basic plot is finally taking shape. But none of the plot points has been settled for good, beginning with a seemingly simple question: what route did we take out of Africa?

Out of Africa – but how?

The route that our ancestors took out of Africa has been repeatedly re-evaluated and is still not firmly agreed.

Based on the evidence of the early occupation of the Middle East, the idea took hold that when early modern humans eventually began their global migration, they took a 'northern route' through the Levant (essentially, the land immediately east of the Mediterranean) and up into Europe. But this has been challenged: other discoveries point to early and widespread occupation of South East Asia and Australasia, with migration to the north and then west into Europe happening later (see Figure 6.2).

For example, in 2007 skeletal remains found in Niah Cave in Sarawak, on the island of Borneo, were dated to between 45,000 and 40,000 years old. Added to these are some fossils from Tianyuan Cave, near Beijing, China, also described in 2007, which are 40,000 years of age.

While some of our ancestors explored the far east of Asia, other groups were beginning to enter Europe. Skeletal remains from Peştera cu Oase, a cave in Romania, also date to about 40,000 years old. The oldest fossils in Western Europe are

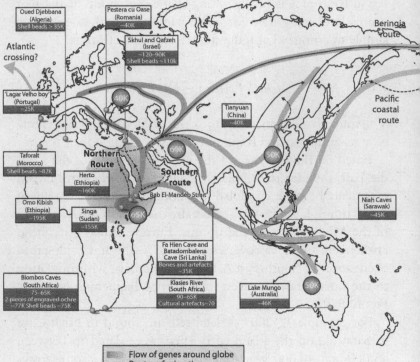

Oued Djebbana (Algeria) Shell beads > 35K

Pestera cu Oase (Romania) ~40K

Skhul and Qafzeh (Israel) ~120–90K Shell beads ~110k

Beringia route

Atlantic crossing?

'Lagar Velho boy' (Portugal) ~25K

Tianyuan (China) ~40K

Pacific coastal route

40K

Taforalt (Morocco) Shell beads ~82K

Northern Route

Herto (Ethiopia) ~160K

60K

Southern route

50K

Omo Kibish (Ethiopia) ~195K

Singa (Sudan) ~155K

Bab El-Mandeb Streit

65K

Niah Caves (Sarawak) ~45K

Fa Hien Cave and Batadombalena Cave (Sri Lanka) Bones and artefacts ~35K

Biombos Caves (South Africa) ~75–65K 2 pieces of engraved ochre ~77K Shell beads ~75K

Klasies River (South Africa) 90–65K Cultural artefacts ~70

Lake Mungo (Australia) ~46K

50K

Flow of genes around globe
Routes of migration
Alternative/contested routes
~13K Anatomically modern humans
10K 10,000 years ago

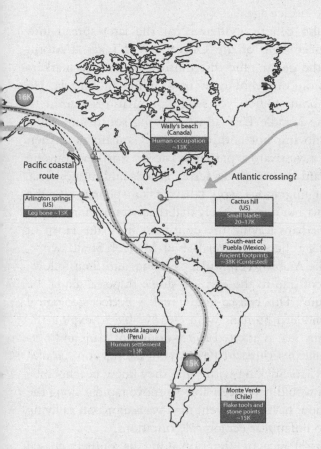

FIGURE 6.2 The migration routes out of Africa of anatomically modern humans. Evidence from fossils, ancient artefacts and genetic analysis combines to tell a compelling story.

slightly younger, between 37,000 and 36,000 years old. Only the Americas seem to have been colonized much later, towards the end of the last ice age.

There is also genetic evidence for the later spread into Europe. Spencer Wells of the University of Texas at Austin has charted the geographic distribution of genetic markers on the Y chromosome of men now living in Eurasia. He found that about 40,000 years ago populations started to diverge in the Middle East, some moving south into India and others moving north through the Caucasus and then splitting into a westward arm that led across northern Europe and an eastward arm reaching across Russia and into Siberia.

These later migrations would have taken people into the heart of Eurasia, but it seems likely that the first migrants skirted the coast – perhaps leaving the continent from the Horn of Africa, crossing a then-narrower Bab el-Mandeb Strait, swinging around the Arabian Peninsula, past Iraq, and then following the coast of Iran to the east – a single dispersal along the 'southern route'. This coastal route makes perfect ecological sense. Early modern humans were clearly able to exploit the resources of the sea, as attested to by dumps of clam and oyster shells found in Eritrea in East Africa, dating from around 125,000 years ago. Sticking with what they knew, beachcombing *H. sapiens* would have been able to move rapidly along the coastline without having to invent new ways of making a living or adapting to unfamiliar ecological conditions.

Archaeological traces of migration along the southern coastal route are patchy but consistent with this picture. But there is also plenty of evidence to support the traditional idea of the northern route. For instance, ancient rivers, whose remains are now lying beneath the Sahara Desert, once formed green corridors at the surface, which our ancestors could have followed on their great trek out of Africa.

Similarly, a 2015 genomic analysis of hundreds of people from the area suggests that we took the northern route through Egypt into Eurasia. This conclusion fits well with other evidence. In particular, Eurasians interbred with Neanderthals soon after they left Africa – and while we know that Neanderthals lingered in the Levant at about the time of the out-of-Africa migration, there is no evidence that they lived further south.

At this point you might be wondering whether, given that there seems to be evidence to support both routes out of Africa, early humans might not simply have followed both. This might accord with the archaeology, but three genetic analyses published in 2016 all suggest that there was only one wave of migration. It seems that all non-Africans living today can trace the vast majority of their ancestry to one group of pioneers who left Africa in a single wave.

An earlier exodus?

Even the date of our departure from Africa is not entirely agreed upon. Our direct ancestors may have found their way out of Africa much earlier than we think.

Jin Changzhu and colleagues of the Institute of Vertebrate Palaeontology and Palaeoanthropology in Beijing announced in 2009 that they had uncovered a 110,000-year-old putative *Homo sapiens* jawbone from a cave in southern China's Guangxi Province. The mandible has a protruding chin like that of *Homo sapiens* but the thickness of the jaw is indicative of more primitive hominins, suggesting that the fossil could derive from interbreeding. Other anthropologists questioned whether the find was a true *Homo sapiens*.

The global spread of modern humans

As always, dates for these events are fluid. There is now evidence of human artefacts in Australia 65,000 years ago, which is older than anything yet found in Asia. That discrepancy has still to be resolved.

At least 200,000 years ago (ya)
Emergence of anatomically modern
humans (AMHs) in Africa

120,000 ya
AMHs briefly occupy Israel
(Skhul and Qafzeh).

42,000 ya
Europe begins to cool further.

60,000 ya
Population explosion in Africa;
AMHs spread across South East Asia.

40,000 ya
Population explosion in Europe;
AMHs in Europe

39,000 ya
Flowering of characteristically modern
behaviour seen in Aurignacian artefacts in
Western Europe, including bone and
antler tools, cave art and jewellery

Last solid evidence of Neanderthals

80,000 ya
Symbolic engravings in ochre and
ornamental beads from marine shells at
Blombos, South Africa

Other shell beads from Taforalt, Morocco

Ice Age begins.

78,000 ya
Population explosion in Africa

65,000 ya
AMHs reach Australia.

70,000 ya
AMHs begin to leave Africa.

18,500 ya
AMHs in Americas

11,000 ya
Ice Age ends.

11,000 ya – today
Period of warm,
stable climate

In 2014 another research group described two teeth from the Luna Cave in China's Guangxi Zhuang region. Based on the proportions of the teeth, the group argued that at least one of them must have belonged to an early *Homo sapiens*. The teeth were clearly old. Calcite crystals, which formed as water flowed over the teeth and the cave floor, date them to between 70,000 and 125,000 years ago. Again, it is not clear that the teeth belong to modern humans.

However, other fossils are more convincing. Bones found in Israel, including an upper jaw from Misliya Cave, could be 150,000 years old. In China, a jawbone and two molars were found in Zhirendong, a cave in Guizhou Province. Although the bone is over 100,000 years old, the shape of its chin is suggestive of modern humans.

Similarly, in 2015 María Martinón-Torres of University College London and her team found 47 teeth, belonging to at least 13 *H. sapiens* individuals, in Fuyan Cave in Daoxian, Hunan Province, south-east China. The teeth were found under a layer of stalagmites that formed after the teeth were deposited there. The stalagmites are at least 80,000 years old, so the teeth seem to be at least that old. Investigation of animal bones at the same site puts their upper age limit at 120,000 years old.

A closer look at the genetics also suggests that there was an earlier migration. In 2014 Katerina Harvati of the University of Tübingen in Germany and her colleagues tested the classic 'out of Africa at 60,000 years ago' story against the earlier-exodus idea. They plugged the genomes of indigenous populations from South East Asia into a migration model. They found that the genetic data was best explained by an early exodus that left Africa 130,000 years ago, taking a coastal route along the Arabian Peninsula, India and into Australia, followed by a later wave along the classic northern route through Egypt.

The conquest of Europe

When the first modern humans entered Europe, they found the Neanderthals already there. What happened next?

Not too long ago, that story appeared beguilingly simple. Our species arrived 45,000 years ago from the Middle East, outcompeted the Neanderthals, and that was that. But since 2010 scientists have used fragments of DNA pulled out of ancient bones to probe Europe's genetic make-up. Together they tell a more detailed, colourful tale: three waves of *H. sapiens* shaped the continent. Each came with its own skills and traits. Together they would lay the foundations for a new civilization.

Our distant ancestors – probably *Homo erectus* – first settled in Europe at least 1.2 million years ago. By 200,000 years ago they had become Neanderthals. We know from their DNA that at least some of them had pale skin and red hair. They lived in caves, had basic stone tools, and hunted and fished along European shorelines. Some probably wore forms of decoration such as shell necklaces and might even have etched simple signs in the rocks of the Iberian Peninsula.

All told, they were relatively sophisticated, but they were no match for a group of dark-skinned hunter-gatherers who arrived from the Middle East around 45,000 years ago. By 39,000 years ago the Neanderthals were history. The newcomers flourished in Europe's forests, hunting the woolly rhinoceros and bison that lived there. Long before farmers tamed cows and sheep, these early Europeans made friends with wolves. Like the Neanderthals, some of them lived at the mouths of caves – sometimes the very same caves – which they lit and heated with roaring fires. Like them, they ate berries, nuts, fish and game.

For 30,000 years the hunter-gatherers had Europe largely to themselves. Then, about 9,000 years ago, Neolithic farmers arrived from the Middle East and began spreading through southern and central Europe. They brought their understanding of how to collect and sow seeds, as well as the staples of modern European diets. The basic livestock of Europe today, with a few exceptions such as chickens, arrived then.

With farming came a more sedentary and recognizably modern lifestyle. Villages became a much more common fixture. Some archaeological sites have distinct rows of houses. A settled lifestyle made it harder to relocate if conflicts arose between neighbouring tribes, so they would often be resolved with mass violence. But, fundamentally, life in early Neolithic Europe was still relatively insular. The final, missing ingredient, the one that would truly lay the flagstones for a European civilization, was still a few thousand years off.

Archaeologists have long debated the existence of another great prehistoric migration into Europe – one that brought in a mysterious group from the Eurasian steppe. The Yamnaya, so the argument went, founded some of the late Neolithic and Copper Age cultures, including the vast Corded Ware culture – named after its distinctive style of pottery – that stretched from the Netherlands to central Russia.

The weight of archaeological opinion had been against this idea, but that may be about to change. A 2014 study of ancient DNA offered the first compelling evidence that a third ancient population did shape the modern European gene pool, and that it originated in northern Eurasia. With the curious exception of the blond hunter-gatherers from Sweden, there were no signs of this population in the genes of early farmers or hunter-gatherers. So the people carrying the genes must have become common in Europe some time after most farmers arrived.

Two separate studies link the arrival of these genes to a massive Yamnaya migration into Europe about 4,500 years ago.

They found that 75 per cent of the genetic markers of skeletons associated with Corded Ware artefacts unearthed in Germany could be traced to Yamnaya bones previously unearthed in Russia. In other words, the genetic testimony provides tantalizing evidence that there really was a massive influx of Yamnaya people into Europe around 4,500 years ago.

The Yamnaya were cattle herders. Until about 5,500 years ago their settlements clung to the river valleys of the Eurasian steppes – the only places where they had easy access to the water they and their livestock needed. The invention of a single, revolutionary technology changed everything. We know from linguistic studies that the Yamnaya rode horses and, crucially, had fully embraced the freedom that came with the wheel.

With wagons, they could take water and food wherever they wanted, and the archaeological record shows that they began to occupy vast territories. This change may have led to fundamental shifts in how Yamnaya society was structured before they left the steppes. Groups roamed into one another's territories, so a political framework emerged that obliged tribal leaders to offer wanderers safe passage and protection. If this is correct, then it is easy to imagine that the Yamnaya brought this new political framework with them as they ventured west.

The land of Oz

How and when people first reached Australia is still being worked out. Until 2017 the oldest known human remains from Australia belonged to a skeleton dubbed Mungo Man, discovered at Lake Mungo in south-eastern Australia in 1974. Experts argued for decades about the age of the skeleton. But in 2003 it was confirmed that Mungo Man lived 40,000 years ago.

That fits with the out-of-Africa story. It also supports the idea that early humans drove Australia's large mammals to extinction. Evidence that people spread rapidly across Australia around 50,000 years ago would support the 'blitzkrieg' theory of the extinction of the continent's large mammals soon after humans arrived. In line with this, a 2016 study suggested that the first humans to reach Australia quickly migrated into its hot, dry interior, and developed tools to adapt to the tough environment and exploit its giant beasts.

However, the story was turned upside down in July 2017, when Chris Clarkson of the University of Queensland in Australia and his colleagues published the results of their excavation of a rock shelter in northern Australia called Madjedbebe. Artefacts from the shelter had previously been dated to 60,000 years ago, but those dates had proved controversial. Clarkson excavated further into Madjedbebe and found artefacts that he could reliably date to 65,000 years ago. As well as simply pushing back the date of Australia's colonization, this called into question the idea that humans were responsible for the extinction of the megafauna – as it now appeared that the two had coexisted for many thousands of years. It also appears that Australia was colonized before the onset of the main migration from Africa 60,000 years ago. It is not clear how that apparent discrepancy will be resolved.

It would be tempting to dismiss Clarkson's findings as a fluke, but a study published just one month later also seems to confound the classic out-of-Africa story. Researchers investigated a cave called Lida Ajer on the Indonesian island of Sumatra, which had previously yielded fossil human teeth. They were able to confirm that the teeth belonged to modern humans, and that they were between 63,000 and 73,000 years old. This is further evidence that modern humans reached Indonesia ahead of schedule – and Australia was then not far away.

There is one other peculiarity about the migration into Australia. The people who first entered the continent seem to have bred with an unknown hominin species. In a 2016 study, researchers analysed the genomes of living indigenous Australians, Papuans, people from the Andaman Islands near India, and from mainland India. They found sections of DNA that did not match any known hominin species.

The New World

The Americas were the last leg of humanity's trek around the globe (if you don't count Antarctica). Not so long ago there was a simple and seemingly incontrovertible answer to the question of how and when the first settlers made it to the Americas. This was that, some 13,000 years ago, a group of people from Asia walked across a land bridge that connected Siberia to Alaska and headed south. These people, known as the Clovis, were accomplished toolmakers and hunters. Subsisting largely on big game killed with their trademark fluted spear points, they prospered and spread out across the continent.

For decades this was the received wisdom. But nowadays the identity of the first Americans is an open question again. According to the Clovis First theory, it was around 13,500 years ago, near the end of the last ice age, that a brief window of opportunity opened up for humans to finally enter North America. With vast amounts of water locked up in ice caps, the sea level was lower than it is today and Siberia and Alaska were connected by a now-submerged land bridge called Beringia, now under the Bering Strait. As the world began to warm, the huge ice sheets that had blocked this way into North America began to retreat, parting to create an ice-free corridor all the way to the east of the Rockies. The Clovis walked right in.

The presence of distinctive stone tools throughout the US and northern Mexico supports the idea that the Clovis lived there 13,000 years ago, as does the timing of an extinction that wiped out more than 30 groups of large mammals including mammoths, camels and sabre-toothed cats. This coincides neatly with the arrival of Clovis hunters, and could have been their handiwork.

But over the years inconvenient bits of evidence have piled up. In 1997 a delegation of 12 eminent archaeologists visited Monte Verde, a site of human habitation in southern Chile that was first excavated in the 1970s. It was originally claimed to be 14,800 years old. That, of course, contradicted the Clovis First theory. The trip was a pivotal moment: most of the visiting archaeologists changed their minds, and prehistory started to be rewritten. Many other pre-Clovis sites have also been found, and in 2015 it emerged that Monte Verde might actually be more than 18,000 years old.

Recent DNA studies also contradict the old orthodoxy. By comparing the genomes of modern Asian and Native American people and estimating the amount of time it would take for the genetic differences to accumulate, geneticists estimate that people entered the Americas at least 16,000 years ago – 3,000 years earlier than in the Clovis model.

So if Clovis First is wrong, how and when were the Americas colonized? Despite the shift, some things stay the same. Most researchers still think that the first Americans migrated from Asia, a conclusion based largely on DNA research. For example, a 2012 study led by Harvard's David Reich compared DNA from Native American populations scattered from the Bering Strait to Tierra del Fuego with DNA from native Siberians. The team concluded controversially that the first Americans were descended from Siberians who arrived in at least three waves (see Figure 6.3).

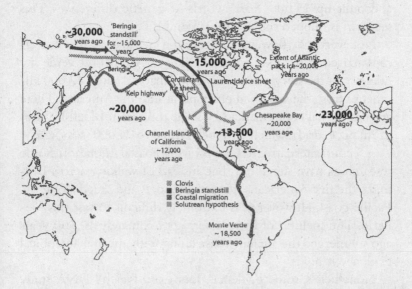

FIGURE 6.3 Alternative migration routes from the Old World to the New are now being considered to explain genetic and archaeological findings.

That might not sound too different from Clovis First. But other research tells another story. The DNA of Native Americans is sufficiently different from the Siberians' DNA to suggest that the populations went their separate ways around 30,000 years ago. This could indicate a much earlier entry. Some archaeologists claim that they have found evidence of human habitation in the Americas as far back as 50,000 years ago, but these claims are contentious.

A more likely scenario is that the settlers did not come directly from Asia but from a population that settled in Beringia 30,000 years ago and stayed put for 15,000 years before pressing on to Alaska. This group would have become isolated from the ancestral population in north-eastern Asia, perhaps by ice, and built up 15,000 years' worth of genetic differences. This scenario is dubbed the 'Beringia standstill'.

Not everybody, however, clings to the idea that the first Americans arrived from Siberia. The 18,000-year-old Monte Verde site does not fit the story, particularly given that it is about 12,000 kilometres from the supposed entry point into the Americas. There are also animal bones marked by stone tools from Bluefish Caves in Yukon, Canada, which have been dated to 24,000 years ago.

A second alternative to Clovis First is coastal migration. Some researchers have suggested that, instead of walking across Beringia, the first Americans hopped on boats and sailed along the Pacific coast. However, it is extremely difficult to test this scenario. The melting of the glaciers approximately 10,000 years ago submerged the ancient coast, along with any archaeological evidence it holds.

Nonetheless, some evidence does exist. Reich's DNA study suggests that the first wave of colonists moved south along the Pacific coast. And we know that ancient East Asians were accomplished seafarers, reaching the isolated Ryukyu Islands between Japan and Taiwan roughly 35,000 years ago. There are also archaeological finds that lend credence to the idea. University of Oregon archaeologist Jon Erlandson has worked on the Channel Islands, west of what is now Los Angeles, for decades and has uncovered evidence of an advanced 12,000-year-old culture there. The remains of 13,000-year-old Arlington Springs Man were also found on the Channel Islands.

Support for coastal migration also comes from Michael Waters, director of Texas A&M University's Center for the Study of the First Americans. He has led the excavation of Friedkin, a pre-Clovis site in central Texas. Since digging began in 2006, over 15,000 artefacts have been uncovered – more than at all other pre-Clovis sites combined – dating from 15,500 to 13,200 years ago. The great majority are offcuts from toolmaking, but there are also choppers, scrapers, hand axes, blades and bladelets. In Waters's opinion, these could be the precursors of Clovis technology. If the dating is right, then these people arrived before the ice sheets retreated and needed some other route south. This adds further credence to a coastal migration.

The Anzick Child

In 2014 geneticists studied the genes of an American boy, known as the Anzick Child, who died 12,600 years ago. He was the earliest ancient American to have their genome sequenced. Incredibly, he turned out to be a direct ancestor of many peoples across the Americas. The find offered the first genetic evidence for what Native Americans have claimed all along: that they are directly descended from the first Americans. It also confirmed that those first Americans can be traced back at least 24,000 years, to a group of early Asians and a group of Europeans who mated near Lake Baikal in Siberia.

We may never know who the Anzick Child was – why he died, at just three years old, in the foothills of the American Rockies; why he was buried, 12,600 years ago, beneath a huge cache of sharpened flints; or why his kin left him with a bone tool that had been passed down the generations for 150 years.

Eske Willerslev of the University of Copenhagen in Denmark and his colleagues were able to extract enough viable DNA from the boy's badly preserved bones to sequence his entire genome. They then compared this with DNA samples from 143 modern non-African populations, including 52 South American, Central American and Canadian tribes.

The comparison revealed a map of ancestry. The Anzick Child is most closely related to modern tribes in Central and South America, and is equally close to all of them – suggesting that his family were common ancestors. To the north, Canadian tribes were very close cousins. DNA comparisons with Siberians, Asians and Europeans show that the further west populations are from Alaska, the less related they are to the boy.

An even earlier entry?

The date of the first entry to the Americas is clearly still in doubt. A few scientists have even claimed evidence for much earlier peopling. In 2003, for instance, 40,000-year-old human footprints were found in volcanic ash near Puebla in Mexico. According to Silvia Gonzalez of Liverpool John Moores University in the UK, this suggested that the New World was colonized far earlier than anyone thought. But later work showed that the ash solidified 1.3 million years ago, long before our species even evolved. Everyone, including Gonzalez, now agrees that the footprints were not really footprints.

Similarly, in 2013 researchers claimed to have found 22,000-year-old stone tools at a site in Brazil. This would imply that humans lived in South America at the height of the last ice age – not as extreme a claim as Gonzalez's, but still a long way from the consensus. But not everyone agrees that the purported tools are real.

The most startling claim of all was published in 2017. Finds at a site in California suggested that the New World might have first been reached at least 130,000 years ago – more than 100,000 years earlier than conventionally thought. If the evidence stacks up, the earliest people to reach the Americas may have been Neanderthals or Denisovans rather than modern humans.

The evidence came from a coastal site in San Diego County. In the early 1990s routine highway excavations exposed fossil bones belonging to a mastodon, an extinct relative of the elephant. Many of the bones and teeth were fragmented. Alongside these remains, the researchers found stone cobbles that had evidence of impact marks on their surfaces, suggesting that humans used stone tools to break the bones. But a dating study indicates that the remains are 131,000 years old.

There were no human fossils at the site. But both Neanderthals and Denisovans were probably present in Siberia more than 130,000 years ago. Since sea levels were low and a land bridge existed between Siberia and North America just before that time, either group could in theory have wandered across. Alternatively, it might have been modern humans – *Homo sapiens* – that made it to the New World 130,000 years ago.

For now, this is all speculation, because most anthropologists do not believe the results. The supposed stone hammers and anvils are a particular weak point because, once again, they might not really be tools.

7
Civilization and beyond

For most of our species' existence, we lived as hunter-gatherers – often in small groups that we might nowadays label 'tribes'. Then, around 10,000 years ago, something changed. We began living in densely populated villages, towns and eventually cities. We worked out how to read and write, make machinery, erect towering buildings and monuments, and we started fighting wars. In short, we invented civilization.

The real first farmers

In February 1910 British botanist Lilian Gibbs walked across North Borneo and climbed Mount Kinabalu, a lone white woman among 400 locals. She later wrote: 'The "untrodden jungle" of fiction seems to be non-existent in this country. Everywhere the forest is well worked and has been so for generations.'

What Gibbs saw was a seemingly curated tropical forest, regularly set alight by local tribes and with space carefully cleared around selected wild fruit trees, to give them room to flourish. The forest appeared to be partitioned and managed to get the most rattan canes, fibre for basketry, medicinal plants and other products. Generation after generation of people had cared for the trees, gradually shaping the forest they lived in. This was not agriculture as we know it today but a more ancient form of cultivation, stretching back more than 10,000 years. Half a world away from the 'Fertile Crescent', the region of the Middle East once thought to be the site of the first settled agricultural communities, Gibbs was witnessing a living relic of the earliest human farming.

In recent years, archaeologists have found signs of this 'proto-farming' on nearly every continent, transforming our picture of the dawn of agriculture. Gone is the simple story of a sudden agricultural revolution in the Fertile Crescent that produced benefits so great that it rapidly spread around the world. It turns out that farming was invented many times, in many places, and was rarely an instant success. In short, there was no agricultural revolution.

Farming is seen as a pivotal invention in the history of humanity. Previously, our ancestors had roamed the landscape gathering edible fruits, seeds and plants and hunting whatever game they could find. They lived in small mobile groups that usually set up temporary homes according to the movement of their prey. Then one fine day in the Fertile Crescent, around 8,000 to 10,000 years ago – or so the story goes – someone noticed sprouts growing out of seeds they had accidentally left on the ground. Over time, people learned how to grow and care for plants to get the most out of them. Doing this for generations gradually transformed the wild plants into rich domestic varieties, most of which we still eat today.

This series of events is credited with irreversibly shaping the course of humanity. As fields began to appear on the landscape, more people could be fed. Human populations – already on the rise and stretching the resources available to hunter-gatherers – exploded. At the same time, our ancestors swapped their migratory habits for sedentary settlements: these were the first villages, with adjoining fields and pastures. A steadier food supply freed up time for new tasks. Craftspeople were born – the first specialized toolmakers, farmers and carers. Complex societies began to develop, as did trade networks between villages. The rest, as they say, is history.

The enormous impact of farming is widely accepted, but in recent decades the story of how it all began has been completely turned on its head. We now know that, while the inhabitants of the Fertile Crescent were undoubtedly some of the earliest farmers, they were not the only ones. Archaeologists now agree that farming was independently 'invented' in at least 11 regions, from Central America all the way to China (see Figure 7.1). Decades of digging have exposed numerous instances of ancient proto-farming, similar to what Gibbs saw in Borneo.

Another point archaeologists are rethinking is the notion that our ancestors were forced into farming when their populations outgrew what the land could provide naturally. If humans

• Fertile Crescent   • New centres of farming

FIGURE 7.1 Farming emerged several times around the world and not only in the Fertile Crescent.

had turned to crops out of hunger and desperation, you would expect their efforts to have intensified when the climate took a turn for the worse. In fact, archaeological sites in Asia and the Americas show that earliest cultivation happened during periods of relatively stable, warm climates when wild foods would have been plentiful.

Nor is there much evidence that early farming coincided with overpopulation. When crops first appeared in eastern North America, for example, people were living in small, scattered settlements. The earliest South American farmers also lived in the very best habitats, where resource shortages would have been least likely. Similarly, in China and the Middle East, domesticated crops appear well before dense human populations would have made foraging impractical.

Instead, the first farmers may have been pulled into experimenting with cultivation, presumably out of curiosity rather than necessity. That lack of pressure would explain why so many societies kept crops as a low-intensity sideline – a hobby, almost – for so many generations. Only much later in the process would densely populated settlements have forced people to abandon wild foods in favour of a near-exclusive reliance on farming.

Those first experiments probably happened when bands of hunter-gatherers started changing the landscape to encourage the most productive habitats. On the islands of South East Asia, people were burning patches of tropical forest as far back as the last ice age. This created clearings where plants with edible tubers could flourish. In Borneo, evidence of this stretches back 53,000 years; in New Guinea it is 20,000 years. We know that the burns were deliberate because the charcoal they left behind peaked during wet periods, when natural fires would be less common and people would be fighting forest encroachment.

Burning forest would have paid off for hunters, too, as game is easier to spot at forest edges. At Niah Cave on the northern coast of Borneo, researchers have found hundreds of orangutan bones among the remains of early hunters, suggesting that forest regrowth after a burn brought the apes low enough to catch, even before the invention of blowpipes. Burning probably intensified as the last ice age gave way to the warmer, wetter Holocene beginning about 13,000 years ago. Rainfall in Borneo doubled, producing a denser forest that would have been much harder to forage without fire.

This wasn't happening only in South East Asia. Changing climates also pushed hunter-gatherers into landscape management in Central and South America. At the end of the last ice age, the perfect open hunting grounds of the savannahs began to give way to closed forest. By 13,000 years ago people were burning forests during the dry season when fires would carry. Researchers are now turning up evidence of similar management activities in Africa, Brazil and North America.

From burning, it is just a short step to actively nurturing favoured wild species, something that also happened soon after the end of the last ice age in some places. Weeds that thrive in cultivated fields appear in the Fertile Crescent at least 13,000 years ago, for example, and New Guinea highlanders were building mounds on swampy ground to grow bananas, yams and taro about 7,000 years ago. In parts of South America, traces of cultivated crops such as gourds, squash, arrowroot and avocado appear as early as 11,000 years ago.

Evidence suggests that these people lived in small groups, often sheltering under rock overhangs or in shallow caves, and that they tended small plots along the banks of seasonal streams in addition to foraging for wild plants. Their early efforts would not have looked much like farming today. They were more like gardens: small, intensively managed plots on riverbanks and alluvial fans that probably did not provide many calories. Instead, they may have provided high-value foods, such as rice, for special occasions.

Archaeologists have long assumed that this proto-farming was a short-lived predecessor to fully domesticated crops. They believed that the first farmers quickly transformed the plants' genetic make-up by selecting traits like larger seeds and easier harvesting to produce modern domestic varieties. After all, similar selection has produced great changes in dogs within just the past few hundred years.

But new archaeological sites and better techniques for recognizing ancient plant remains have made it clear that crop domestication was often very slow. Through much of the

Middle East, Asia and New Guinea, at least a thousand – and often several thousand – years of proto-farming preceded the first genetic hints of domestication.

And in China people began cultivating wild forms of rice on a small scale about 10,000 years ago. But physical traits associated with domesticated rice, such as larger grains that stay in the seed head instead of falling off to seed the next generation, did not appear until about two-and-a-half millennia later. Fully domesticated rice did not appear until 6,000 years ago.

Even after crops were domesticated, there was often a lag, sometimes of thousands of years, before people began to rely on them for most of their calories. During this prolonged transition period people often acted as though they had not decided how much to trust the new-fangled agricultural technology.

The record also shows a long period of overlap in other regions, with cultures using both wild foods and domesticated crops. We know from the type of starch grains found on their teeth that people living in southern Mexico 8,700 years ago were eating domesticated maize, yet large-scale slash-and-burn agriculture did not begin until nearly a millennium later. In several cases – Scandinavia, for example – societies began to rely on domesticated crops and then switched back to wild foods when they could not make a success of farming. And in eastern North America, Native Americans had domesticated squash, sunflowers and several other plants by about 3,800 years ago, but only truly committed to agriculture about 1,100 years ago.

The story of agriculture, in short, is not the sudden agricultural revolution of textbooks but, rather, an agricultural evolution. It was a long, drawn-out process.

How we lived as hunter-gatherers: an interview with Jared Diamond

Jared Diamond is Professor of Geography at the University of California, Los Angeles. Here he discusses his research into today's tribal communities to see how we all once lived, for his 2012 book The World Until Yesterday.

What distinguishes tribal communities' way of child-rearing?

Outside observers are universally struck by the precocious social skills of children in tribal societies. In most traditional cultures, kids have the right to make their own decisions. Sometimes this horrifies us because a two-year-old can decide to play next to a fire and burn itself. But the attitude is that children have got to learn from their own experience.

What are the roles of family and community in these traditional societies?

Children sleep with their parents so they have absolute security, and are nursed whenever they want. They live in multi-age playgroups, so, by the time they are teenagers, they have spent ten years bringing up little siblings.

Are there any negative aspects?

There are many things that these societies do that are wonderful, and there are some things that they do that, to us, seem terrible – like occasionally killing their old people or infants, or persistently making war.

In what context would elderly people be killed?

In most traditional societies that live in permanent villages, old people have happier and more satisfying lives than those in Western society. They spend their lives with their relatives, children and lifelong friends. And in a society without writing, old people are valued for their insight and knowledge. But in a nomadic society, the cool reality is that, if you have to move and you are already carrying your baby and your stuff, you can't carry the old person. There's no alternative.

What happens to old people in nomadic tribes?

There's a sequence of choices that all end up with them being abandoned or killed. The gentlest thing to do is to abandon them, leaving food and water in case they regain strength and can catch up. In some societies, old people take it upon themselves to ask to be killed. In other societies, they are killed actively. Among the Aché people of Paraguay, for example, there are young men whose job it is to kill the older people.

The first civilizations

Many archaeologists also suspect that farming was not, as long believed, the first step on the road to civilization.

Steven Mithen at the University of Reading, UK, spent years digging for Stone Age ruins in south Jordan. He found the remains of a primitive village. As the team dug through what they thought was a rubbish dump, a student came upon a polished, solid floor – hardly the kind of craftsmanship to waste on a communal tip. Then came a series of platforms engraved with wavy symbols. It was clearly not the local dump.

Mithen now compares the structure to a small amphitheatre. With benches lining one side of a roughly circular building, it looks purpose built for celebrations or spectacles – perhaps feasting, music, rituals or something more macabre. A series of gullies runs down through the floor, through which sacrificial blood might once have flowed in front of a frenzied crowd. Whatever happened at the place now known as Wadi Faynan, the site could transform our understanding of the past. At 11,600 years old, it predates farming – which means that people were building amphitheatres before they invented agriculture.

It was not supposed to be that way. Archaeologists have long been familiar with the idea of a 'Neolithic Revolution' during which humans abandoned the nomadic lifestyle that had served them so well for millennia and settled in permanent agrarian communities. They domesticated plants and animals and invented a new way of life.

By about 8,300 years ago people in the Levant – modern-day Syria, Lebanon, Jordan, Israel, the Palestinian territories and parts of southern Anatolia in Turkey – had the full range of Neolithic technologies. settled villages with communal buildings, pottery, domesticated animals, cereals and legumes. Art, politics and astronomy also have their roots in this time. And yet here was a settlement more than 3,000 years older displaying many of those innovations, but lacking the technology that is supposed to have got the whole thing started: farming. The people who built Wadi Faynan were not nomads but neither were they farmers. They probably relied almost exclusively on hunting and gathering.

Instead of agriculture, then, some very different motivations seem to have drawn these people together – things like religion, culture and feasting. Never mind the practical benefits of a steady food supply; the seeds of civilization may have been sown by something much more cerebral.

For much of the twentieth century our view of the Neolithic was seen through the lens of more recent social upheaval: the Industrial Revolution. The idea originated, in part, with Marxist archaeologist Vere Gordon Childe. Seeing the urban societies that had coalesced around factory towers and 'dark satanic mills', Childe suspected that the first farms could have been similar hotbeds of rapid social and cultural change.

He proposed that it began in the Levant around 10,000 years ago. As the Ice Age ended, the region became more arid, save for smaller patches of lush land by rivers. With these limited areas to forage, nomadic hunter-gatherers discovered that it was more efficient to cultivate barley and wheat in one place. A baby boom followed. As Childe put it in his 1936 book *Man Makes Himself*: 'If there are more mouths to feed, there will also be more hands to till the fields ... quite young toddlers can help in weeding fields and scaring off birds.' And as the farmers' crops and families blossomed, so, too, did their crafts, including carpentry and pottery, along with greater social complexity. The growing communities would have also been fertile ground for more organized forms of religion to flourish.

At least, that was the theory. *Man Makes Himself* became a touchstone for many archaeologists – even as cracks began to appear in some of its assumptions. Studies of the climate, for instance, suggest that the changes following the Ice Age were not nearly as radical as Childe believed. Without the environmental spark, there were doubts that agriculture offered any real benefits. Particularly when you have only a few bellies to fill, plundering nature's larder is just as efficient as the backbreaking business of planting, weeding and harvesting. So why would you change?

By the 1990s those cracks had turned to gaping chasms, following digs in Anatolia. The region was already attracting attention for a site known as Nevali Çori, which was around 10,000 years old. Although it seemed to be a simple settlement of proto-farmers, the archaeologists also uncovered signs of more advanced culture, embodied in a series of communal 'cult buildings' full of macabre artwork.

The buildings were remarkably large and complex for something so old. And what they contained was even more revealing. In one sculpture, a snake writhes across a man's head; another depicts a bird of prey landing on the heads of entwined twins. The most eye-catching feature was a collection of strange, anthropomorphic T-shaped megaliths with faceless, oblong heads and human arms engraved on their sides. As people sat on benches around the walls of the buildings, these monuments must have loomed over them like sentinels.

Sadly, the site was submerged when the Atatürk Dam was built across the Euphrates. But one of the archaeologists, Klaus Schmidt, set about scouring the surrounding countryside for further clues to the origins of this lost society. During this tour he found himself on a mound called Göbekli Tepe. The grassy knoll was already popular with locals visiting its magic 'wishing tree', but what really caught Schmidt's eye was a large piece of limestone that closely resembled those T-shaped megaliths from Nevali Çori (see Figure 7.2).

It did not take him long to realize that he had stumbled on something even more extraordinary. Buried beneath the hill, he found three layers of remains. The oldest and most impressive was more than 11,000 years old, with a labyrinth of circular 'sanctuaries' measuring up to 30 metres in diameter. Around the inner walls were magnificent, T-shaped monuments encircling two larger pillars, like worshippers surrounding their

FIGURE 7.2 Two of the T-shaped monuments of Göbekli Tepe. What
were they for?

idol. Some were engraved with belts and robes and, given their monumental size – around three times the height of a modern man – and abstract appearance, Schmidt interprets them as representing some kind of god-like figure.

If Nevali Çori was a humble parish church, then this was a cathedral. Strangely, each sanctuary seems to have been dismantled and deliberately filled in some time later – perhaps as part of a ritual. Amid the jumble of debris, Schmidt's team have found many bones, including human remains. His team has also found a lot of rooks and crows – birds known to be drawn to corpses. For this reason, Schmidt's team believe that some of the buildings' functions may have centred on death.

We can never know what happened there, but Schmidt has some suspicions. From the outset, he was fascinated by strange door-like 'porthole stones', found within the sanctuaries and often decorated with grisly images of predators and prey. Since the holes in the middle are often the size of a human body, Schmidt imagines that visitors may have crawled through to symbolize the passage into the afterlife.

It is clear that Göbekli Tepe was the creation of a sophisticated society, capable of marshalling the labour of perhaps hundreds of people. That degree of social complexity was not expected in emerging early Neolithic cultures. Along with the complex artwork and intricate ideology, this kind of development was supposed to come long after agriculture. Yet Schmidt failed to find any signs of farming. Domesticated corn can be distinguished from its wild ancestor by its plumper ears, but there was no trace of it. Stranger still, there is no sure evidence of any kind of permanent settlement at Göbekli Tepe. It was too far away from water supplies and Schmidt found little evidence of the hearths, fire pits or tools you might expect in a dwelling.

Schmidt's conclusions were radical. He proposed that Göbekli Tepe was a dedicated site of pilgrimage, perhaps the culmination of a long tradition of gatherings and celebrations. Importantly, it was ideology rather than farming that was pulling these people together to form a larger society. Indeed, it may have been the need to feed people at these kinds of gatherings that eventually led to agriculture – which turns the original idea of the Neolithic revolution on its head. Tellingly, recent genetic work pinpoints the origin of domestic wheat to a spot very close to Göbekli Tepe.

Schmidt's finds astonished archaeologists and captivated the wider world. The 'first temple' soon began attracting a new swarm of pilgrims, with filmmakers, archaeologists and tourists flocking to visit.

Other researchers are dubious. The original people's habit of periodically burying their sanctuaries means that there is always the possibility that old remains rather than contemporary debris were dug up to dump on the monuments. That would shave hundreds or thousands of years off the age of the temple, making it much less revolutionary. Others doubt Schmidt's claims that Göbekli Tepe was the site of pilgrimage rather than a permanent settlement.

Such concerns do not necessarily derail Schmidt's broader theory that culture, rather than farming, propelled our march to civilization. But it was clear that, to expand the theory, archaeologists needed to look further afield. Fortunately, they were on the trail almost as soon as Göbekli Tepe was discovered. A little down the Euphrates, across the border into Syria, French researchers have found a trio of early Neolithic villages called Dja'De, Tell'Abr and Jerf el-Ahmar. Although they are clearly permanent settlements rather than sites of pilgrimage, they all house large, highly decorated communal buildings that seem to have been the product of the same complex, ritualistic culture as Göbekli Tepe (see Figure 7.3).

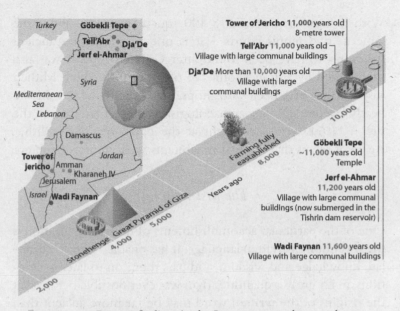

FIGURE 7.3 Recent findings in the Levant suggest that people were living in large settlements and building temples before they invented agriculture.

With Syria's civil war raging, these settlements are now off limits, but charred remains of seeds caught in cooking pots and house fires at Jerf el-Ahmar revealed that the first inhabitants were still gathering a wide variety of wild cereals and lentils. Later on, however, in the upper layers, a few species begin to dominate – ones that would later be domesticated. There is also evidence of imported crops that would not naturally grow in the region. So the people of Jerf el-Ahmar were probably cultivating plants by the latter stages of its occupation. The main point, though, is that they had begun to build their complex society long before they had domestic crops.

The 'amphitheatre' at Wadi Faynan, Jordan, which Mithen first excavated in 2010, tells a similar story much further south.

With a floor area of nearly 400 square metres – about the same as two tennis courts – it is one of the largest ancient structures to have been found after Göbekli Tepe. It was also surrounded by a 'honeycomb' of other rooms, which Mithen suspects may have been workshops. Importantly, the remains are neatly layered, allowing archaeologists to pin a firm date on the site – 11,600 years ago, right at the dawn of the Neolithic. Again, it seems that the first inhabitants were hunter-gatherers.

The first writing

One of the particular accomplishments of human civilization is the invention of written language. It has enabled us to accumulate knowledge and wisdom, and pass them on to later generations, in far greater quantities than was ever possible before. But the origins of the written word may be far more ancient than we ever suspected.

Genevieve von Petzinger, a palaeoanthropologist from the University of Victoria in Canada, is leading an unusual study of cave art. Her interest lies not in the breathtaking paintings of bulls, horses and bison that usually spring to mind, but in the smaller, geometric symbols frequently found alongside them. Her work has convinced her that, far from being random doodles, the simple shapes represent a fundamental shift in our ancestors' mental skills.

The first formal writing system that we know of is the 5,000-year-old cuneiform script of the ancient city of Uruk in what is now Iraq. But it and other systems like it – such as Egyptian hieroglyphs – are complex and did not emerge from a vacuum. There must have been an earlier time when people first started playing with simple abstract signs. For years, von Petzinger has wondered whether the circles, triangles and squiggles that humans began leaving on cave walls 40,000 years ago represent that special time in our history – the creation of the first human code.

If so, the marks are highly significant. Our ability to represent a concept with an abstract sign is something no other animal, not even our closest cousins the chimpanzees, can do. It is arguably also the foundation for our advanced, global culture.

Between 2013 and 2014 von Petzinger visited 52 caves in France, Spain, Italy and Portugal. The symbols she found ranged from dots, lines, triangles, squares and zigzags to more complex forms like ladder shapes, hand stencils, something called a tectiform that looks a bit like a post with a roof, and feather shapes called penniforms. In some places, the signs were part of bigger paintings. Elsewhere, they were on their own, like the row of bell shapes found in El Castillo in northern Spain or the panel of 15 penniforms in Santian, also in Spain.

Thanks to von Petzinger's meticulous logging, it is now possible to see trends – new signs appearing in one region and remaining common for a while before falling out of fashion. The research also reveals that modern humans were using two-thirds of these signs when they first settled in Europe, which creates another intriguing possibility. 'This does not look like the start-up phase of a brand new invention', von Petzinger wrote in her book *The First Signs* (2016). In other words, when modern humans first started moving into Europe from Africa, they must have brought a mental dictionary of symbols with them.

That fits well with the discovery of a 70,000-year-old block of ochre etched with cross-hatching in Blombos Cave in South Africa. And when von Petzinger looked through archaeology papers for mentions or illustrations of symbols in cave art outside Europe, she found that many of her 32 signs were used around the world (see Figure 7.4). There is even tantalizing evidence that an earlier human, *Homo erectus*, deliberately etched a zigzag on a shell on Java some 500,000 years ago.

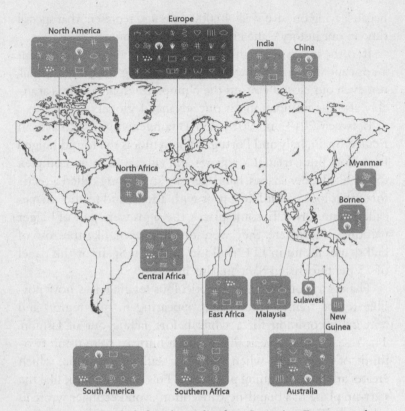

FIGURE 7.4 Symbols found on relics from Stone Age Europe are also found in caves across the rest of the world. Their similarities suggest that they were more than just random squiggles.

Nonetheless, something quite special seems to have happened in Ice Age Europe. In various caves, von Petzinger frequently found certain symbols used together. For instance, starting 40,000 years ago, hand stencils are often found alongside dots. Later, between 28,000 and 22,000 years ago, they are joined by thumb stencils and finger fluting – parallel lines created by dragging fingers through soft cave deposits.

These kinds of combinations are particularly interesting if you are looking for the deep origins of writing systems. Nowadays, we effortlessly combine letters to make words and words to make sentences, but this is a sophisticated skill. Von Petzinger wonders whether the people of the Upper Palaeolithic, which began 40,000 years ago, started experimenting with more complex ways of encoding information using deliberate, repeated sequences of symbols. Unfortunately, it is hard to say from signs painted on cave walls, where arrangements could be deliberate or completely random.

Whether or not the symbols are actually writing depends on what you mean by 'writing'. Strictly speaking, a full system must encode all of human speech, ruling the Stone Age signs out. But if you take it to mean a system to encode and transmit information, then it is possible to see the symbols as early steps in the development of writing.

Von Petzinger has another reason to believe that the symbols are special: they are easy to draw. In a sense, the humble nature of such shapes makes them more universally accessible – an important feature for an effective communication system. More than anything, she believes that the invention of the first code represents a complete shift in how our ancestors shared information. For the first time, they no longer had to be in the same place at the same time to communicate with one another, and information could survive its owners.

Did palaeo-porn exist? An interview with April Nowell

April Nowell is a Palaeolithic archaeologist. In 2014 she published a paper, co-authored with Melanie Chang, exploring whether certain prehistoric figurines might have been pornographic.

Which Palaeolithic images and artefacts have been described as pornography?

The Venus figurines of women, some with exaggerated anatomical features, and ancient rock art, like the image from the Abri Castanet site in France that is supposedly of female genitalia.

You take issue with this interpretation. Who is responsible for spreading it, journalists or scientists?

People are fascinated by prehistory, and the media want to write stories that attract readers – and, to use a cliché, sex sells. But when a *New York Times* headline reads 'A Precursor to *Playboy*: Graphic Images in Rock' and *Discover* magazine asserts that man's obsession with pornography dates back to 'Cro-Magnon days' based on 'the famous 26,000-year-old Venus of Willendorf statuette … [with] GG-cup breasts and a hippopotamal butt', I think a line is crossed. To be fair, archaeologists are partially responsible – we need to choose our words carefully.

Aren't other interpretations of palaeo-art just as speculative as calling them pornographic?

Yes, but when we interpret Palaeolithic art more broadly, we talk about 'hunting magic' or 'religion' or 'fertility magic'. I don't think these interpretations have the same social ramifications as pornography. When respected journals – *Nature*, for example – use terms such as 'prehistoric pin-up' and '35,000-year-old sex object', and a German museum proclaims that a figurine is either an 'earth mother or pin-up girl' (as if no other roles for women could have existed in prehistory), they carry weight and authority. This allows journalists and researchers, evolutionary psychologists in particular, to legitimize and naturalize contemporary Western values and behaviours by tracing them back to the 'mist of prehistory'.

Shaped by war

War has bedevilled human societies throughout recorded history. The accounts of the earliest civilizations, such as the Sumerians and Mycenaean Greeks, are replete with battles – and archaeologists have uncovered evidence of battles and massacres going back even further. All of this raises the question: why do we go to war? The cost to human society is enormous and yet, for all our intellectual development, we continue to wage war well into the twenty-first century.

Since the early 2000s a new theory has emerged that challenges the prevailing view that warfare is a product of human culture and thus a relatively recent phenomenon. For the first time, anthropologists, archaeologists, primatologists, psychologists and political scientists have approached a consensus. Not only is war as ancient as humankind, they say, but it has also played an integral role in our evolution. The theory helps explain the evolution of familiar aspects of warlike behaviour such as gang warfare. It even suggests that the cooperative skills we have had to develop to be effective warriors have turned into the modern ability to work for a common goal.

Mark Van Vugt, an evolutionary psychologist at the Free University Amsterdam in the Netherlands, thinks warfare was already there in the common ancestor we share with chimps, and that it has affected our evolution. Studies suggest that warfare accounts for 10 per cent or more of all male deaths in present-day hunter-gatherers.

Primatologists have known for some time that organized, lethal violence is common between groups of chimpanzees, our closest relatives. Whether between chimps or hunter-gatherers, however, intergroup violence is nothing like modern pitched battles. Instead, it tends to take the form of brief raids using overwhelming force, so that the aggressors run little risk of injury. This opportunistic violence helps the aggressors weaken rival groups and thus expand their territorial holdings.

Such raids are thought to be possible because humans and chimps, unlike most social mammals, often wander away from the main group to forage singly or in smaller groups. Bonobos – which are as closely related to humans as chimps are – have little or no intergroup violence, perhaps because they tend to live in habitats where food is easier to come by, so they need not stray from the group.

If group violence has been around for a long time in human society, then we ought to have evolved psychological adaptations to a warlike lifestyle. In line with that, there is evidence that males – whose larger and more muscular bodies make them better suited for fighting – have evolved a tendency towards aggression outside the group, but cooperation within it.

Aggression in women, by contrast, tends to take the form of verbal rather than physical violence, and is mostly one-to-one. Gang instincts may have evolved in women, too, but to a much lesser extent. This may be partly because of our evolutionary history, in which men are often stronger and better suited for physical violence. This could explain why female gangs tend to form only in same-sex environments such as prisons or high schools. But women also have more to lose from aggression, since they bear most of the effort of child-rearing.

Not surprisingly, a study has shown that men are more aggressive than women when playing the leader of a fictitious country in a role-playing game. But there were also more subtle responses in group bonding. For example, male undergraduates were more willing than females to contribute money towards a group effort – but only when competing against rival universities. If told instead that the experiment was to test their individual responses to group cooperation, men coughed up less cash than women did. In other words, men's cooperative behaviour emerged only in the context of intergroup competition.

Some of this behaviour could arguably be attributed to conscious mental strategies, but anthropologist Mark Flinn of the University of Missouri at Columbia has found that group-oriented responses occur on the hormonal level, too. He found that cricket players on the Caribbean island of Dominica experience a testosterone surge after winning against another village. But this hormonal surge, and presumably the dominant behaviour it prompts, was absent when the men beat a team from their own village. Similarly, the testosterone surge a man often has in the presence of a potential mate is muted if the woman is in a relationship with his friend. Again, the effect is to reduce competition within the group.

The net effect of all this is that groups of males take on their own special dynamic. Think of soldiers in a platoon or football fans out on the town: cohesive, confident and aggressive – just the traits a group of warriors needs. Chimpanzees don't go to war in the way we do because they lack the abstract thought required to see themselves as part of a collective that expands beyond their immediate associates.

However, it may be that the real story of our evolutionary past is not simply that warfare drove the evolution of social behaviour. According to Samuel Bowles, an economist at the Santa Fe Institute in New Mexico and the University of Siena, Italy, the real driver was 'some interplay between warfare and the alternative benefits of peace'. Although women seem to help broker harmony within groups, men may be better at peacekeeping between groups.

Our warlike past may have given us other gifts as well. In particular, warfare requires a colossal level of cooperation. And that seems to be a heritage worth hanging on to.

The most peaceful time: an interview with Steven Pinker

Steven Pinker is Professor of Psychology at Harvard University. His 2011 book The Better Angels of Our Nature *argued that humans are less violent now than at any point in history.*

Where did you find evidence for how violence has changed over time?

For prehistoric times, the main evidence is from forensic archaeology: the proportion of skeletons that had bashed-in skulls or arrowheads embedded in bones, together with archaeological evidence such as fortifications. For homicide over the last millennia or so, there are records in many parts of Europe that go back to the Middle Ages. And we know from documents of the era that crucifixions and all manner of gory executions took place in the ancient world.

For data about wars, there are many databases that estimate war deaths, and in recent eras governments and social scientists have tracked just about every aspect of life, so we really can get a clear view of things like child abuse, spousal abuse, rape and so on.

How do you explain the decline in violence?

I don't think there is a single answer. One cause is government, that is, third-party dispute resolution: courts and police with a monopoly on the legitimate use of force. Everywhere you look for comparisons of life under anarchy and life under government, life under government is less violent. The evidence includes transitions such as the European homicide decline since the Middle Ages, which coincided with the expansion and consolidation of kingdoms and the transition from tribal anarchy to the first states. Watching the movie in reverse, in today's failed states violence goes through the roof.

Do you think commerce helps, too?

Commerce, trade and exchange make other people more valuable alive than dead, and mean that people try to anticipate what the other guy needs and wants. It engages the mechanisms of reciprocal altruism, as the evolutionary biologists call it, as opposed to raw dominance.

What else has contributed to the decline?

The expansion of literacy, journalism, history, science – all of the ways in which we see the world from the other guy's point of view. Feminization is another reason for the decline. As women are empowered, violence can come down, for a number of reasons. By all measures men are the more violent gender.

The evolution of religion

The rise of human civilization has been accompanied by another peculiar phenomenon: religious belief, and all the ceremonies, monuments and moral certainties that go with it. How did religion – which, on the face of it, is irrational and costly – survive and even thrive? Let's start with a thought experiment.

By the time he was five years old Wolfgang Amadeus Mozart could play the clavier and had begun to compose his own music. Mozart was a 'born musician'; he had strong natural talents and required only minimal exposure to music to become fluent. Few of us are quite so lucky. Music usually has to be drummed into us by teaching, repetition and practice. And yet in other domains, such as language or walking, virtually everyone is a natural; we are all 'born speakers' and 'born walkers'.

So what about religion? Is it more like music or language? Evidence from developmental psychology and cognitive anthropology indicates that religion comes to us nearly as naturally as language: we just need minimal exposure and our natural predilections do the rest. The vast majority of humans are 'born believers', naturally inclined to find religious claims and explanations attractive and easily acquired, and to attain fluency in using them. This attraction to religion is an evolutionary by-product of our ordinary cognitive equipment and human sociality.

For example, children learn early the difference between objects and 'agents' that act on those objects. They learn that agents, even if they cannot be seen, make things happen and that agents often act for a purpose. Experiments with children support the idea that they expect apparent order and design in the world to be brought about by an agent. This makes children naturally receptive to the idea that there may be one or more gods, which helps to account for the world around them.

It is important to note that this concept of religion deviates from theological beliefs. Children are born believers, not of Christianity, Islam or any other theology, but of what we might call 'natural religion'. They have strong natural tendencies towards religion, but these tendencies do not inevitably propel them towards any one religious belief. Instead, the way the human mind solves problems generates a god-shaped conceptual space, waiting to be filled by the details of the culture into which we are born.

Religion could even help explain a deep puzzle of civilization: how did human societies scale up from small, mobile groups of hunter-gatherers to large, sedentary societies? The puzzle is one of cooperation. Up until about 12,000 years ago all humans lived in relatively small bands. Today, virtually everyone lives in vast, cooperative groups of mostly unrelated strangers. How did this happen?

In evolutionary biology, cooperation is usually explained by one of two forms of altruism: cooperation among kin and reciprocal altruism – you scratch my back and I'll scratch yours. But cooperation among strangers is not easily explained by either of these. As group size increases, both forms of altruism break down. With ever-greater chances of encountering strangers, opportunities for cooperation among kin decline. Reciprocal altruism – without extra safeguards such as institutions for punishing freeloaders – also rapidly stops paying off.

This is where religion comes in. Some early cultural variants of religion presumably promoted pro-social behaviours such as cooperation, trust and self-sacrifice, while encouraging displays of religious devotion, such as fasts, food taboos, extravagant rituals and other 'hard-to-fake' behaviours that reliably transmitted believers' sincere faith and signalled their intention to cooperate. Religion thus forged anonymous strangers into moral communities tied together with sacred bonds under a common supernatural jurisdiction.

In turn, such groups would have been larger and more cooperative, and hence more successful in competition for resources and habitats. As these ever-expanding groups grew, they took their religions with them, further increasing social solidarity in a runaway process that softened the limitations on group size imposed by kinship and reciprocity.

From there, it is a short step to the morally concerned 'Big Gods' of the major world religions. People steeped in the Abrahamic faiths are so accustomed to seeing a link between religion and morality that it is hard for them to imagine that religion did not start that way. Yet the gods of the smallest hunter-gatherer groups, such as the Hadza of East Africa and the San of the Kalahari, are unconcerned with human morality. In these transparent societies, where face-to-face interaction is the norm, it is hard to escape the social spotlight. Kin altruism and reciprocity are sufficient to maintain social bonds.

However, as groups expand in size, anonymity invades relationships and cooperation breaks down. Studies show that feelings of anonymity – even illusory ones, such as wearing dark glasses – are the friends of selfishness and cheating. Social surveillance, such as being in front of a camera or an audience, has the opposite effect. Even subtle exposure to drawings resembling eyes encourages good behaviour towards strangers. As the saying goes, 'Watched people are nice people.'

It follows, then, that people play nice when they think a god is watching them, and those around them. The anthropological record supports this idea. In moving from the smallest-scale human societies to the largest and most complex, Big Gods – powerful, omniscient, interventionist watchers – become increasingly common, and morality and religion become increasingly intertwined.

As societies get larger and more complex, rituals become routine and are used to transmit and reinforce doctrines. Similarly, notions of supernatural punishment, karma, damnation and salvation, and heaven and hell are common in modern religions, but relatively infrequent in hunter-gatherer cultures.

Religion, with its belief in watchful gods and extravagant rituals and practices, has been a social glue for most of human history. But recently some societies have succeeded in sustaining cooperation with secular institutions such as courts, police and mechanisms for enforcing contracts. In some parts of the world, especially Scandinavia, these institutions have precipitated religion's decline by usurping its community-building functions. These societies with atheist majorities – some of the most cooperative, peaceful and prosperous in the world – have climbed religion's ladder and then kicked it away.

Nine more things that made us human

1 Weapons

Projectile weapons travel faster than even the speediest antelope. A study published in 2013 suggests that *H. erectus* made use of them, since it was the earliest of our ancestors with a shoulder suitable for powerful and accurate throwing. What's more, unusual collections of fist-sized rocks at the *H. erectus* site at Dmanisi, Georgia, give us an idea of their projectile weapon of choice.

But throwing rocks did more than offer a new hunting strategy: it also gave early humans an effective way to kill an adversary. Christopher Boehm at the University of Southern California says that projectile weapons levelled the playing field in early human societies, by allowing even the weakest group member to take down a dominant figure without having to resort to hand-to-hand combat. Weapons, he argues, encouraged early human groups to embrace an egalitarian existence unique among primates – one that is still seen in hunter-gatherer societies today.

2 Jewellery and cosmetics

At the Blombos Cave in South Africa, excavations in the early 2000s revealed collections of shells that had been perforated and stained, then strung together to form necklaces or bracelets. Similar finds have now turned up at other sites in Africa. More recently, work at Blombos has uncovered evidence that ochre was deliberately collected, combined with other ingredients and fashioned into body paint or cosmetics.

At first glance, these inventions seem trivial, but they hint at dramatic revolutions in the nature of human beliefs and communication. Jewellery and cosmetics were probably prestigious, suggesting the existence of people of higher and lower status. More importantly, they are indications of symbolic thought and behaviour, because wearing a particular necklace or form of body paint has meaning beyond the apparent. As well as status, it can signify things like group identity or a shared outlook. That generation after generation adorned themselves in this way indicates that these people had language complex enough to establish traditions.

3 Sewing

What people invented to wear with their jewellery and cosmetics was equally revolutionary. Needle-like objects appear in the archaeological record about 60,000 years ago, providing the first evidence of tailoring, but humans had probably already been wearing simple clothes for thousands of years.

Evidence for this comes from a rather unusual source. Body lice, which live mostly in clothes, evolved from hair lice some time after humans began clothing themselves, and a 2003 study of louse genetics suggests that body lice arose some 70,000 years ago. A 2011 analysis puts their origin as early as 170,000 years ago. Either way, it looks as if we were wearing sewn clothes when we migrated from our African cradle some 60,000 years ago and began spreading across the world.

Clothes may have allowed humans to inhabit cold areas that their naked predecessors could not tolerate. Sewing could have been a crucial development, since fitted garments are more effective at retaining body heat than loose animal furs. Even then, the frozen north would have been a challenge for a species that evolved on the African savannah, and recent research indicates that we also took advantage of changes in the climate to move across the world.

4 Containers

When some of our ancestors left Africa, they probably travelled with more than just the clothes on their backs. About 100,000 years ago, people in southern Africa began using ostrich eggs as water bottles. Having containers to transport and store vital resources would have given them huge advantages over other primates. But engravings on these shells are also highly significant: they appear to be a sign that dispersed groups had begun to connect and trade.

Since 1999 Pierre-Jean Texier at the University of Bordeaux in France has been uncovering engraved ostrich egg fragments at the Diepkloof rock shelter, 150 kilometres north of Cape Town in South Africa. The same five basic motifs are used time and again, over thousands of years, implying that they had a meaning that numerous generations could read and understand. Texier and his colleagues think they show that people were visually marking and defining their belongings, maintaining their group identity as they began travelling further and interacting with other groups.

5 **Law**

As our ancestors began trading, they would have needed to cooperate fairly and peacefully – with not just group members but also strangers from foreign lands. Trade may have provided the impetus to invent law and justice to make everyone follow the same rules.

Hints of how law evolved come from modern human groups, which, like Stone Age hunter-gatherers, live in egalitarian, decentralized societies. The Turkana are nomadic pastoralists in East Africa. Despite having no centralized political power, the men will cooperate with non-family members in a life-threatening venture – stealing livestock from neighbouring peoples, say. While the activity itself may be ethically dubious, the motivation to cooperate reflects ideas that underpin any modern justice system. If men refuse to join these raiding parties, they are judged harshly and punished by other group members. The Turkana display mechanisms of adjudication and punishment akin to a formal judiciary, suggesting that law and justice predate the emergence of centralized societies.

6 **Timekeeping**

As trade flourished over the millennia that followed, it was not just material goods that were exchanged. Trade in ideas encouraged new ways of thinking, and perhaps the early stirrings of scientific thought. Communities of hunter-gatherers living in what is now Scotland may have been among the first to scientifically observe and measure their environment. Aberdeenshire has many Mesolithic sites dating from about 10,000 years ago, including an odd monument consisting of a dozen pits arranged in a shallow arc roughly from north-east to south-west. When Vincent Gaffney at the University of Bradford and his colleagues noticed that the arc faced a sharp valley on the horizon through which the sun rises on the winter solstice, they realized that it was a cosmological statement. The 12 pits were almost certainly used to keep track of lunar months. The Aberdeenshire lunar 'calendar' – or 'time reckoner', as they dubbed it – is comfortably twice the age of any previously found.

By establishing a formal concept of time, you know when to expect seasonal events, such as the return of salmon to the local rivers. And such knowledge is power.

7 Ploughing

While Scotland's hunter-gatherers were measuring time, their contemporaries in the Near East had settled down to farm. Crop cultivation is tough work that inspired the first farmers to invent labour-saving devices. The most iconic of these, the plough, might have influenced society in a surprising way.

In the past, as today, hunter-gatherer societies were probably often divided along gender lines, with men hunting and women gathering. Farming promised greater gender equality, because both sexes could work the land, but the plough – which was heavy and so primarily controlled by men – brought an end to that. So argued Danish agricultural economist Ester Boserup in the 1970s. In 2013 Paola Giuliano at the University of California, Los Angeles, and her colleagues tested the idea by comparing gender equality in societies across the world that either adopted the plough or a different form of agriculture. Not only did they confirm the plough effect, they found that it continues to influence gender perceptions today.

8 Sewerage

Farming has been described as the worst mistake in human history: it is back-breaking work. But it did provide such plentiful food that it allowed the growth of urban centres. City living comes with many advantages, but it also carries a health warning: urbanites are at risk from infectious diseases carried by water.

Almost as long as there have been cities, there have been impressive sewerage systems. Cities in the 5,000-year-old Indus Valley society were built above extensive drains. Lavatory-like systems existed in early Scottish settlements dating from around the same time, and there are 3,500-year-old flush toilets and sewers in Crete. But none of these were really designed with sanitation in mind. They were really just to dispose of waste water, often into the nearest river.

It was only in the 1850s, after physician John Snow linked an outbreak of cholera in London to insanitary water supplies, that people started to clean waste water. Large-scale centralized sewage works date from the early decades of the twentieth century. Effective sewerage was a long time coming, but when it did arrive it revolutionized public health.

9 Writing

The engraved ostrich eggshells of Diepkloof show that modern humans have used graphical symbols to convey meaning for at least 100,000 years. But genuine writing was invented only about 5,000 years ago. Ideas could spread and cultural evolution would never be the same again.

Writing also provided a means to convey hopes and fears, revealing how subsequent innovations had affected the human psyche. Some of the world's oldest texts, from the Mesopotamian city-state of Lagash, rail against the spiralling taxes exacted by a corrupt ruling class. Soon afterwards, King Urukagina of Lagash wrote what is thought to be the first documented legal code. He has gained a reputation as the earliest social reformer, creating laws to limit the excesses of the rich, for instance. But his decrees also entrench the inferior social position of women. One details penalties for adulterous women, but makes no mention of adulterous men. Despite all our revolutionary changes, humanity still had some way to go.

8
A special species

Clearly, humans are unique. We are the only species in the Earth's 4.5-billion-year history to have invented advanced technologies like writing, computers, rockets and fidget spinners. We have also changed the world around us: there is a good case that we have started a new geological era called the Anthropocene, defined by our influence on the world. But what is so special about us? Is it just our powerful brains, or are there other factors that have made us (at least for now) so successful?

The cultured ape

Homo sapiens is the only species with a history. In the relatively brief span of our existence, we have gone from upright apes with a few hand-axes and spears to a species that spread from Africa to occupy nearly every habitat on Earth, building a world replete with technologies most of us do not even understand. By contrast, our close genetic cousins, chimpanzees, still sit on the ground cracking nuts with stones, as they have for millions of years. History for other animals really is, as British historian Arnold Toynbee said, 'just one damn thing after another' – and the same 'thing' at that.

Our achievements pose a challenge to Darwin. His great theory of evolution by natural selection provides a sophisticated view of how species adapt to their environments. But how are we to explain the existence of petrol engines, cameras, pasta machines, yo-yos, religion and the arts? Even if we concoct stories to explain how these artefacts might improve our survival, why have only humans produced them? What was it about humans that set us on a trajectory of cumulative and accelerating technological innovation, the limits of which we are still exploring?

We know it must have been small in genetic terms because we share around 98 per cent of our protein-coding gene sequences with chimpanzees, and more than 99 per cent with the hapless and extinct Neanderthals. And yet there seems to be an unbridgeable gap between our evolutionary potential and theirs. Indeed, there seems to be a gap between us and all other species.

The usual human conceit is that we are simply more intelligent: our big brains allow us to figure things out. But this view is exaggerated. Looking at the evolution of technology, it did not happen with great leaps of insight, but through small and often accidental modifications to existing ideas. Thomas Edison's notebooks show that he tried thousands of materials, including platinum and bamboo, before alighting on a carbon fibre as the filament for his light bulb; Henry Ford didn't invent the assembly line; even Isaac Newton acknowledged that he stood on the shoulders of giants.

Instead, it may be that what our species is really good at is imitation. We can search among a sea of what might be little more than random ideas others have tried, picking the ones that seem to work best. It is a kind of 'survival of the fittest ideas' that mimics biological evolution, and because ideas can quickly spread from one mind to another, the pace of our cultural evolution vastly outstrips the plodding rate of most genetic change.

However, there is more to this story. Copying is fraught with errors. If left uncorrected, those errors will accumulate on top of other errors, and this will eventually bring the cultural evolutionary train to a halt, at least for things more complicated than those you might be able to learn on your own. This is a fair description of most other animals' technologies – chimpanzees, for example, probably rediscover the art of nut cracking every generation, perhaps benefiting only from having their attention called to it from watching others. Lacking a mechanism to reduce copying errors, the chimpanzees are stuck at this level of sophistication.

Our solution may have been to teach. Teaching can transmit new information, but it is also an error-correction mechanism that allows more sophisticated practices and technologies to be passed on and accumulate. Some animals do display rudimentary forms of teaching – such as when adult meerkats disable the stinger on a scorpion to allow their offspring to experiment with it at low risk – but only humans practise the systematic teaching of complex actions. It has even been suggested that our human capacity for language evolved, not for the economic and social reasons many others suggest, but as an aid to teaching: language arose as something akin to an aural DNA.

Our cultural adaptations have equipped us for the modern world, but they have also left legacies. Today, the advance of culture and the changes it has wrought in us have yielded a species that is curiously in and out of its time.

What big brains you have

Let us look at perhaps our most obvious unique attribute: our brains are enormous relative to our body size. While it is worth bearing in mind that size isn't everything (the design and connectivity of the brain are also crucial, for example), it seems reasonable to assume that our big brains helped to make us clever.

With a volume of 1,200 to 1,500 cubic centimetres, our brains are three times the size of those of our nearest relatives, the chimpanzees. This expansion may have involved a kind of snowball effect, in which initial mutations caused changes that were not only beneficial in themselves but also allowed subsequent mutations that enhanced the brain still further.

In comparison to that of a chimp, the human brain has a hugely expanded cortex, the folded outermost layer that is home to our most sophisticated mental processes, such as planning, reasoning and language abilities. One approach to finding the genes involved in brain expansion has been to investigate the causes of primary microcephaly, a condition in which babies are born with a brain one-third of the normal size, with the cortex particularly undersized. People with microcephaly are usually cognitively impaired to varying degrees.

Genetic studies of families affected by primary microcephaly have so far turned up seven genes that can cause the condition when mutated. Intriguingly, all seven play a role in cell division, the process by which immature neurons multiply in the foetal brain before migrating to their final location. In theory, if a single mutation appeared that caused immature neurons to undergo just one extra cycle of cell division, that could double the final size of the cortex.

Take the gene *ASPM*, short for 'abnormal spindle-like microcephaly-associated'. It encodes a protein found in immature neurons that is part of the spindle – a molecular scaffold that shares out the chromosomes during cell division. We know that this gene was undergoing major changes just as our ancestors' brains were rapidly expanding. When the human *ASPM* sequence was compared with that of seven primates and six other mammals, it showed several hallmarks of rapid evolution since our ancestors split from chimpanzees.

Other insights come from comparing the human and chimp genomes to pin down which regions have been evolving the fastest. This process has highlighted a region called HAR1, short for 'human-accelerated region-1', which is 118 DNA base pairs long. We do not yet know what HAR1 does, but we do know that it is switched on in the foetal brain between 7 and 19 weeks of gestation, in the cells that go on to form the cortex. Equally promising is the discovery of two duplications of a gene called *SRGAP2*, which affect the brain's development in the womb in two ways: the migration of neurons from their site of production to their final location is accelerated; and the neurons extrude more spines, which allow neural connections to form.

Going back a little further in time, a single mutation may have cleared the way for rapid brain evolution. Other primates have strong jaw muscles that exert a force across the whole skull, constraining its growth. But around 2 million years ago a mutation weakened this grip in the human line. A brain growth spurt began soon after.

While it is tough to work out just how our brains got so big, one thing is certain: all that thinking requires extra energy. The brain uses about 20 per cent of our energy at rest, compared with about 8 per cent for other primates. The transition to eating meat would have helped. So would the addition of seafood about 2 million years ago, providing omega-3 fatty acids for brain building. Cooking might have helped too, by making digestion easier. This would have allowed ancestral humans to evolve smaller guts and devote the spare resources to brain building (see Chapter 4).

What is more, several mutations have been discovered that may have helped to power the brain. One emerged with the publication of the gorilla genome in 2012. This revealed a DNA region that underwent accelerated evolution in an ancient primate ancestor, common to humans, chimps and gorillas, some time between 15 and 10 million years ago.

The region was within a gene called *RNF213*, the site of a mutation that causes Moyamoya disease – a condition that involves narrowing of the arteries to the brain. This suggests that the gene may have played a role in boosting the brain's blood supply during our evolution. Since damaging the gene can affect blood flow, perhaps other changes might influence that in a beneficial way.

There are more ways to boost the brain's energy supply than just replumbing its blood vessels, though. The organ's main food source is glucose and this is drawn into the brain by a glucose-transporter-molecule in the blood-vessel walls. Compared with chimpanzees, orang-utans and macaques, humans have slightly different 'on switches' for two genes that encode the glucose transporters for brain and muscle, respectively. The mutations give us more glucose transporters in our brain capillaries and fewer in our muscle capillaries. In short, it looks as if athleticism has been sacrificed for intelligence.

So much for genes: might our behavioural and lifestyle choices have influenced our brains? One popular idea is the 'social brain hypothesis'. It claims that bigger brains and advanced cognitive abilities are primarily an adaptation to greater social complexity, with natural selection strongly favouring individuals who can outsmart rivals. The main support for this idea has been evidence that primates in bigger groups have larger brains. Large brains in humans supposedly followed from our ancestors living in relatively large groups, according to this argument.

However, a study in 2017 challenged the foundational evidence for this link. Researchers collected measures of primate brains and sociality from more than 140 primate species, about three times as many as before, and found no correlation between indicators of brain size and measures of sociality such as group size. Instead, they found that, among primates, brain volume is correlated with diet, with fruit-eating primates having the biggest brains. If this is correct, our unique cognitive ability was made possible by our uniquely high-quality diets.

Getting naked

One unique fact about humans is that we have hardly any fur. But unlike our big brains, where the benefits are obvious, losing our fur is downright puzzling. Mammals expend huge amounts of energy just keeping warm. A pelt is nature's insulation. Why would we forgo that benefit?

One idea is that we lost our fur when overheating became a risk. This wouldn't have been a problem in the shady forest, but when our ancestors moved to more open ground, natural selection would have favoured individuals with very fine hair that helped cooling air to circulate around their sweaty bodies. But sweating requires a large fluid intake, which means living near rivers or steams, whose banks tend to be wooded and shady – thus reducing the need to sweat. What's more, the Pleistocene ice age set in around 1.6 million years ago and even in Africa the nights would have been chilly.

Besides, other animals on the savannah have hung on to their fur. It may be that we did not shed our pelts until we were smart enough to deal with the consequences, which was probably after modern humans evolved, at least 200,000 years ago. Modern humans can make things to compensate for fur loss, such as clothing, shelter and fire. Then, perhaps, natural selection favoured less hairy individuals because fur harbours parasites that spread disease. Later, sexual selection lent a hand, as people with clear, unblemished skin advertising their good health became the most desirable sexual partners and passed on more genes.

To confuse things further, circumstantial evidence points to a very early denuding. The pubic louse evolved around 3.3 million years ago, and it could not have done so until ancestral humans lost their body fur, creating its niche. What is more, scientists have dated the evolution of body lice, which live in clothing, to between 170,000 and 70,000 years ago. So it looks as though our ancestors wandered around stark naked for a very long time.

On speaking terms

Without language, we would struggle to exchange ideas or to influence other people's behaviour. Human society as we know it could not exist. The origin of this singular skill was a turning point in our history, yet the timing is difficult to pin down.

Homo sapiens was not the only hominin with linguistic abilities. Neanderthals had the neural connections to the tongue, diaphragm and chest muscles necessary to articulate intricate sounds and control breathing for speech. Evidence for this comes from the size of holes in the skull and vertebrae through which the nerves serving these areas pass. Neanderthals also shared the human variant of the *FOXP2* gene, crucial for forming the motor memories involved in speech (see Chapter 5). If this variant arose just once, speech predates the divergence of the human and Neanderthal lines around 500,000 years ago.

Indeed, it appears that *Homo heidelbergensis* could already speak 600,000 years ago when it first appeared in Europe. Fossilized remains show that they had lost a balloon-like organ connected to the voice box that allows other primates to produce loud, booming noises to impress their opponents. On the face of it, that was a big disadvantage, but it may be that those air sacs would have blurred differences between vowels, making it hard to form distinct words.

Indisputable evidence of speech conveying complex ideas comes only with the cultural sophistication and symbolism that is associated with *Homo sapiens*. But the first words, whenever they were spoken, started a chain of events that changed our relationships, society and technology, and even the way we think.

While there is ambiguity about when language arose during hominin evolution, there is one hard data point available:

chimpanzees, our closest living relatives, do not have it. Bring up a chimpanzee from birth as if it were a human and it will learn many unsimian behaviours, like wearing clothes and even eating with a knife and fork. But one thing it will not do is talk. In fact, it would be physically impossible for a chimp to talk like us, thanks to differences in our voice boxes and nasal cavities. There are neurological differences too, some of which result from changes to our *FOXP2* gene.

The story of this 'language gene' began with a British family that had 16 members over three generations with severe speech difficulties. Usually, speech problems are part of a broad spectrum of learning difficulties, but the 'KE' family, as they came to be known, seemed to have deficits that were more specific. Their speech was unintelligible and they had a hard time understanding others' speech, particularly when applying rules of grammar. They also had problems making complex movements of the mouth and tongue.

In 2001 the problem was pinned on a mutation in *FOXP2*. We can tell from its structure that this gene helps regulate the activity of other genes. Unfortunately, we do not yet know which ones are controlled by *FOXP2*. What we do know is that in mice (and so, presumably, in humans) *FOXP2* is active in the brain during embryonic development. It may be that, in humans, *FOXP2* plays a role in how we learn the rules of speech – that specific vocal movements generate certain sounds, perhaps, or even the rules of grammar.

That begins to explain how language evolved, but it does not tell us why. Large brains, the ability to make complex hand gestures, distinctive vocal tracts and *FOXP2* may have sharpened our linguistic skills. However, these traits on their own do not explain why we evolved language. There are animals with larger brains, gesturing is widespread among primates, and some bird species can imitate human speech without our descended larynx or our particular version of *FOXP2*.

Instead, the feature that most clearly separates us from other animals is the sophistication of our symbolic and cooperative social behaviour. Humans are the only species that routinely exchanges favours, goods and services with others outside their immediate family. We have an elaborate division of labour: we specialize at tasks and then trade our products with others. And we have learned to act in coordinated ways outside the family unit, such as when a nation goes to war or people combine their efforts to build a bridge.

We take the complexity of our social behaviour for granted, but all these actions rest on our ability to negotiate, bargain, reach agreements and hold people to them. This requires a conduit – like a modern USB cable – to carry complex information back and forth between individuals. Language is that conduit.

Some social insects – ants, bees and wasps – have a level of cooperation without language. But they tend to belong to highly related family groups, genetically programmed to act largely for the good of the group. Human societies must police anyone who tries to take advantage. With words and symbols, we can expose them as cheats and tarnish their reputations. We can lavish praise on those worthy of it, whose reputations will be elevated even among those they have never met: words can travel farther than a single action.

All these complicated social acts require more than the grunts, chirrups, odours, colours and roars of the rest of the animal kingdom. They tell us why we and we alone have language: our particular brand of sociality could not exist without it.

The kindness paradox

So far, we have discussed how our evolution supposedly endowed us with new and impressive abilities, like intelligence or language. But a crucial factor in our development may be something altogether less definable: kindness. To understand this, imagine that you are a Maasai (if you are a Maasai, this should be easy).

Life isn't easy for a Maasai herder on the Serengeti Plain in East Africa. At any moment, disease could sweep through your livestock, the source of almost all your wealth. Drought could parch your pastures, or bandits could steal the herd. No matter how careful you are or how hard you work, fate could leave you destitute. What's a herder to do?

The answer is simple: ask for help. Thanks to a Maasai tradition known as *osotua* – literally, umbilical cord – anyone in need can request aid from their network of friends. Those who are asked are obliged to help, often by giving livestock, as long as it does not jeopardize their own survival. No one expects a recipient to repay the gift, and no one keeps track of how often a person asks or gives.

Osotua runs counter to the way we usually view cooperation, which is all about reciprocity: you scratch my back and I'll scratch yours. Yet similar forms of generosity turn out to be common in cultures around the world. Some anthropologists think they could represent some of the earliest forms of generosity in human society.

Artistic apes

Another uniquely human phenomenon, which again can seem a little vague, is art. Explaining the peculiar human urge to create works of art in terms of evolutionary survival is a challenge. Darwin suggested that art has its origins in sexual selection, and Geoffrey Miller at the University of New Mexico in Albuquerque has run with the idea. He thinks that art is like a peacock's tail – a costly display of evolutionary fitness.

Miller's studies show that both general intelligence and the personality trait of being open to new experiences correlate with artistic creativity. He has also found that when women are at their monthly peak in fertility, they prefer creative over wealthy men. But still, sex alone may not explain the evolution of art. Perhaps it originated for some other function, and acquired the sexual display function later. So what other purpose might art serve?

One suggestion is that the drive to seek out aesthetic experiences could have evolved to push us to learn about different aspects of the world – those that our brain's hardwiring has not equipped us to deal with at birth. Another idea is that art is a social adaptation. It is all about making an object or event 'special' by appealing to the emotions through, say, colour or rhythm. This process could have helped increase our ancestors' chance of survival by bonding a group together.

However, none of this explains where our aesthetic sense comes from. Michael Gazzaniga of the University of California, Santa Barbara, has suggested that we could be biologically primed to find certain images, such as symmetrical designs, more aesthetically pleasing – more beautiful – simply because our brain can process them more quickly. However, he adds that these days we respond positively to some art not because it appeals to us aesthetically, but because seeing it or, better still, owning it is an indicator of status. This point rather brings us full circle back to Miller's ideas. After all, it takes a good deal of counter-intuitive education to distinguish 'good' from 'bad' contemporary art. Most people don't have the time to acquire such 'elite' aesthetic taste – and that is itself a form of fitness display.

If we are unsure where our artistic bent came from, we are equally uncertain about when art first arose. This story is changing all the time as new discoveries are made and old ones rein-terpreted. What does seem clear is that art is much older than we once thought (see Figure 8.1). We have already come across the 'creative explosion' – a rapid proliferation of cave art and sym-bolic artefacts like jewellery and sculpture that began 50,000 to 40,000 years ago (see Chapter 6). The transition was once thought to be a sign of a sudden cognitive change – perhaps the result of genetic mutations that swept through the human population and ultimately resulted in the modern mind.

The cognitive leap theory has always had its detractors, though, since it seemed to occur well after our ancestors left Africa at least 70,000 years ago. If the breakthrough was indeed caused by a mutation in Europe, how did the genetic change filter through to populations in Australia, Asia or the Americas, that had long since lost contact with their European relatives? A far simpler explanation was that our common ancestors had already evolved the necessary brainpower before leaving Africa – but the evidence was lacking.

This all changed with a series of intriguing finds at the Blombos Cave, South Africa. Featuring artefacts such as ostrich shell beads and blocks of ochre etched with geometric shapes, the site seemed to show signs of symbolic art 30,000 years earlier and 10,000 kilometres further south than the artists who fuelled the 'creative explosion' in Europe. Drawing on these early finds, in

FIGURE 8.1 These cave paintings at Lascaux in France are estimated to be up to 20,000 years old. What made our ancestors start to adorn their world with imagery?

2000 anthropologists Sally McBrearty and Alison Brooks wrote a powerful attack on the 'Eurocentric' view of human origins, in a paper entitled 'The Revolution That Wasn't'.

Their strong criticisms spurred others to look far and wide for the origins of symbolic thought, and many old finds have since been reappraised and new ones uncovered. Among the more notable discoveries are the engraved ostrich egg shells from the Diepkloof rock shelter in South Africa that are at least 52,000 years old (see Chapter 7). Collections of seashells in Qafzeh Cave, Israel, and the Grotte des Pigeons in Morocco, meanwhile, show that modern humans were already collecting personal ornaments 80,000 years ago. And in one of the few finds in Asia, some jewellery from Zhoukoudian Upper Cave near Beijing, China, may be 34,000 years old, again suggesting that groups were experimenting with different ways of communicating and decorating themselves right across the world.

A particularly remarkable discovery was announced in 2014. Archaeologists exploring a limestone cave on the Indonesian island of Sulawesi found a hand stencil next to a drawing of a female babirusa, known as a pig-deer. The babirusa image is at least 35,400 years old, which means that it is among the earliest identified figurative paintings in the world. The hand stencil is at least 39,900 years old, making it the oldest example of this common ancient art form ever found. The images were undoubtedly made by early modern humans, contradicting the idea that artistic expression began with early Europeans.

As finds like these have pushed the emergence of abstract thought deeper and deeper into our evolutionary past, some archaeologists are even questioning whether art and symbolism are unique to *Homo sapiens*. After all, Neanderthals had roughly similar-sized brains to modern humans – and the lack of available evidence so far might just be down to the nature of those species' cultures.

There are certainly some tantalizing hints that Neanderthals may have experimented with art. The El Castillo Cave in Spain contains one such hint. Littered with red dots, outlines of bison and handprints, it is quite a spectacle, so it is easy to miss the most remarkable artefact – a red painted disc, almost completely masked by an opaque layer of calcite. An analysis of the uranium content of the calcite suggests that it was painted at least 40,000 years ago, making it the oldest evidence of painting to have been identified in Europe. Since modern humans were only just arriving in Europe at the time, some researchers have concluded that it could have Neanderthal origins.

Along similar lines, researchers have found blocks of manganese pigments in caves at the Pech de l'Azé site in France, which were occupied by Neanderthals. Shaped like crayons, the blocks might have been used to paint designs on the body – in itself a symbolic act.

It is even possible that more distant relatives were budding artists. In Israel, for instance, researchers unearthed the 230,000-year-old Berekhat Ram figurine, which appears to resemble what are known as Venus figurines carved in Europe some 30,000 years ago. The statue is crude and could be nothing more than a conveniently shaped pebble, although some microscopic analyses indicate that there was deliberate carving around the neck to sculpt it into the right proportions. If so, the timing and location suggest that it was the handiwork of *H. erectus*. Naturally, this idea remains contentious.

Whatever the conclusions on other species' abilities, it seems clear that the capacity for abstract thought had arisen long before our ancestors had left Africa. But that does not solve the mystery of the creative explosion, with its figurative art and mythical creatures. What caused our ancestors to take the leap from those early efforts to the intricate creations found in Europe's caves?

One possibility is that their populations had reached a critical mass that somehow promoted innovation – an idea supported by recent findings showing that *H. sapiens* experienced a population explosion once they reached Europe. Advanced cultures, after all, are the product of many smaller innovations – and that needs many minds thinking and inventing over many years.

Beyond the brain

Big brains may have allowed the intellect, language skills and creativity of our ancient ancestors to flourish, but being human is not just about what's between our ears. Nobody would mistake a human for a chimpanzee, for example. Yet we share more DNA with chimps than mice do with rats. Line up the genomes of humans and chimps side by side and they differ by little more than 1 per cent. That may not seem like much, but it equates to more than 30 million point mutations. Around 80 per cent of our genes are affected, and although most have just one or two changes, these can have dramatic effects. The mutation of the human *FOXP2* gene is just one example.

But protein evolution is only part of what makes us human. Also critical are changes in gene regulation – when and where genes are expressed during development. Mutations in key developmental genes are likely to be fatal. But altering the expression of a gene in a single tissue or at a single time can more easily lead to an innovation that is not lethal. Many scientists are comparing gene expression in different tissues to home in on the key regulatory difference between chimps and humans, most of which have still to be uncovered.

Then there is gene duplication. This can give rise to families of genes that diversify and take on new functions. Evan Eichler at the University of Washington in Seattle has identified uniquely human gene families that affect many aspects of our biology, from the immune system to brain development. He suspects that gene duplication has contributed to the evolution of novel cognitive capacities in humans, but at a cost: greater susceptibility to neurological disorders.

Copying errors mean that whole chunks of DNA have been accidentally deleted. Other chunks find themselves in new locations when mobile genetic elements jump around the genome or viruses integrate themselves into our DNA. The human genome contains more than 26,000 of these 'INDELs', many linked with differences in gene expression between humans and chimps.

One key feature of humans is our skilful hands. From the first simple stone tools, through to the control of fire and the development of writing, our progress has been dependent on our dexterity. Can our DNA shed light on our unrivalled abilities with tools? Clues come from a DNA region called HACNS1, short for 'human-accelerated conserved non-coding sequence 1', which has undergone 16 mutations since we split from chimps. The region is an on/off switch that seems to kick a gene into action in several places in the embryo, including developing limbs. Cutting and pasting the human version of HACNS1 into mouse embryos reveals that the mutated version is activated more strongly in the forepaw, right in the areas that correspond to the human wrist and thumb.

Some speculate that these mutations contributed to the evolution of our opposable thumbs, which are crucial for the deft movements required for tool use. In fact, chimps also have opposable thumbs, but just not to the same extent as us.

9
Shaped by our world

The imagery of human evolution sometimes gives the impression of a thrusting, determined species, labouring to become a success regardless of what was happening in the world. But, in fact, humans and our cousins are just like other species: we are part of a wider ecosystem; and the course of our evolution has been affected by the places we have lived and the species with which we have coexisted. If the world had been different, we would be different; we might not even be here at all.

The changing climate

One of the most hotly debated notions, discussed for a century, is that the Earth's ever-changing climate might have somehow affected human evolution. One persistent idea is that the challenging climate of southern Africa – a sparsely vegetated, dry savannah – drove hominins to walk on two legs, grow large brains and develop technology.

But, by the 1990s, Rick Potts from the Smithsonian Institute in Washington, DC had a new theory. He had concluded that the critical part of the human evolutionary story is that our lineage became extremely versatile, capable of living in all kinds of habitats. We are not masters of savannah life but master invaders. This led Potts to suggest that maybe it was environmental change itself – not a particular environment – that drove human evolution. A variable climate, he argued, would have placed a premium on being nimble and versatile.

A climate that shifts from wet to dry every 10,000–20,000 years would have selected for humans with a capacity to adjust to change, whatever that change might be. For example, big brains would have allowed us to solve problems caused by changes in rainfall, such as being able to make different stone tools to exploit changing food resources.

In 1996 Potts published the idea in his book *Humanity's Descent*, calling it 'variability selection'. In 2015 a series of papers by Potts and others finally presented evidence. What had been missing before was a clear link between periods of highly variable climate and milestones in human evolution, so, over two decades, Potts and others gathered evidence of past climates at sites where early humans lived. This allowed them to pinpoint periods of highly variable and stable climate at five such sites in Africa, from between about 3.5 and 1 million years ago.

They then modelled the distribution of key events – things like the appearance or demise of a hominin species, migrations and the development of new technologies – over the past 5 million years, to see what you would expect if climate variability were not driving human evolution. They could then compare this with the actual distribution of events.

From chance alone, the team calculated that you would expect to see five speciation events overlap with periods of high climate variability. Their findings showed that eight overlapped (see Figure 9.1). Similarly, chance alone would predict about four of the seven technological shifts overlapping with climatic variability, but they found that six did.

However, it is not clear that any of these major changes in humans were actually a result of natural selection. It is just as likely that what Potts attributes to natural selection might actually be other types of evolution, such as random genetic drift, which would mean that climate did not have an important role at all. These uncertainties notwithstanding, the correlations between human evolution and climate change keep piling up.

One such piece of evidence comes from East Africa's Turkana Basin. Part of the Great Rift Valley straddling Kenya and Ethiopia (see Figure 9.2), Turkana has yielded many crucial hominin fossils. But was Turkana a recurring site for major events in human evolution, perhaps because it was a humid refuge for our ancestors during particularly dry periods, or was it simply a good environment for preserving fossils?

East Africa as a whole became drier between 3 and 2 million years ago – the period when our genus, *Homo*, first emerged. But the Turkana Basin began to dry out earlier, which means that it could have acted as a 'species factory'. New species that evolved there were adapted for the drier environment that later became widespread. They would have been effectively 'ahead of the trend'.

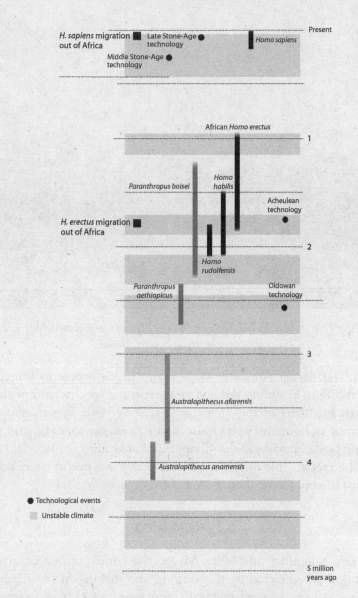

FIGURE 9.1 Chance alone cannot account for the close association between changing climate and significant events in hominin evolution.

FIGURE 9.2 Was the environment of the African Great Rift Valley perfect for creating resilient new species?

The drying coincides with many major events in human evolution, including the appearance in the fossil record of the first members of the *Homo* genus, along with *Paranthropus*, a group of hominins known for their robust skeletons and grinding teeth. *Australopithecus* disappeared at around the same time. The specific role of the climate shift in these events is unclear, but it would have changed what foods were available.

Plate tectonics

Here is another possibility. Maybe the course of human evolution was directed by the shifting and shaking of the Earth's crust. Living in very dynamic landscapes could have selected for adaptability, much as a variable climate might have.

Archaeologist Geoff Bailey of the University of York in the UK and Geoffrey King of the Paris Institute of Earth Physics in France have spent decades amassing evidence for the theory. They argue that our ancestors evolved into modern humans while inhabiting tectonically active regions. Intelligent species would have thrived in these deformed landscapes, exploiting the topography to hunt, avoid predators and competitors, and build defensible homes. Eventually, they developed large brains, a prolonged childhood and the use of advanced tools and weapons. Less smart species would not have had the ability to use the uneven ground to their advantage.

The Earth's surface is divided into plates, which move around over the millennia. Where they grind against each other, pressure builds up, and this can trigger earthquakes and volcanic eruptions. But Bailey and King are concerned with subtler effects. In active regions, the folding and faulting of the crust, combined with regular earthquakes and volcanic activity, create a disrupted landscape with many hills, valleys and cliffs, crisscrossed with solidified volcanic lava.

Bailey argues that these complex landscapes were perfect for early humans, who were not fast runners or particularly strong, but intelligent and adaptable. For instance, even though weapons such as spears had not been invented, early hunters could kill large animals by exploiting the irregularity of the landscape. What's more, since humans evolved from tree-living primates, they would have found it easy to switch to clambering around hills and valleys. By contrast, they would have been at a disadvantage on flat, open plains like the African savannah, which is dominated by fast-running predators like lions and hyenas.

Tectonically active landscapes are also more likely to have reliable water sources, because earthquakes can trap water behind barriers of rock, forming lakes, and underground water can rise through faults to form springs. These water supplies would support plants and attract animals. Barriers like cliffs and ridges would have made life safer by allowing early humans to hide from predators and defend themselves against invaders.

If this theory is right, we should find that early humans were clustered in tectonically active regions. In a 2010 study Bailey and King superimposed the locations of human fossil sites throughout Africa with satellite images that show the roughness of the land, and found that they lined up neatly. In fact, 93 per cent of the fossil sites are in regions of high or medium surface roughness. For example, most of the classic human fossil sites, like Olduvai Gorge and Laetoli, are found along the East African Rift, where two continental plates are slowly coming apart. Bailey and King have found similar patterns of fossil sites in Arabia, which humans colonized later.

However, that left a big problem. Buried remains are more likely to be thrown up on to the surface if they are in an earthquake-prone region, so the results could be misleading. To get around this, Bailey and King, with Sally Reynolds of Bournemouth University, UK, extended their studies to South Africa, where human remains have been found in sites including Taung and Makapansgat. Rather than having been revealed by tectonic activity, the remains were found in caves. There are hundreds of such caves, but only some have remains and those are in regions that were tectonically active, they found.

In a further study, published in 2015, Bailey, King and others analysed archaeological sites in Europe and Asia. Once again, corridors of complex, hilly landscapes lined up well with the distribution of early human fossils.

Supervolcanoes

Toba is a supervolcano on the Indonesian island of Suma-tra. It has blown its top many times but one eruption, 74,000 years ago, was exceptional. Releasing 2,500 cubic kilometres of magma – nearly twice the volume of Mount Everest – the eruption was more than 5,000 times as large as the 1980 erup-tion of Mount St Helens in the USA, making it the largest eruption on Earth in the past 2 million years. The disaster is particularly significant since it occurred at a crucial period in human prehistory – when Neanderthals and other hominins roamed much of Asia and Europe, and around the time that our direct ancestors, *Homo sapiens*, were first leaving Africa to ultimately conquer the world.

Our ideas about this eruption have changed a great deal since 2000. Previous computer models had suggested that the event was truly cataclysmic – very nearly a doomsday for early humankind. With calculations based on the assumption that Toba belched out 100 times more aerosols than the 1991 eruption of Mount Pinatubo in the Philippines, and scaling the environmental effects accordingly, the models suggested that global temperatures dropped by about 10 °C following the blast. This supports the idea of a decade-long 'volcanic winter' and widespread catastrophe.

To make matters worse, the aerosols would have blocked out life-giving sunlight and absorbed water vapour in the atmos-phere, causing a dryer global climate for the next few years. This would have resulted in a rapid decline in tree cover and a concomitant expansion of grasslands, leading to the extinction of many mammals and nearly wiping out our ancestors.

Indeed, the event may have drastically altered the path of evolution for our own species, *Homo sapiens*. Modern humans, who were still thought to be living in Africa, would have been reduced to just a few thousand breeding pairs scattered in dispersed refugia – creating a 'genetic bottleneck' in evolution. As the separate colonies developed independently of one another, they would have sown the seeds for the genetic differences between races once these separate groups eventually left Africa.

This theory has drawn criticism in recent years. Scholars such as Hans Graf, an atmospheric scientist at the University of Cambridge, now believe that the climate change following the explosion has been wildly overestimated. He and his colleagues have suggested a new estimate of global cooling of 2.5 °C lasting just a few years. According to this model, the effects were also highly regional. In places like India, the average temperatures may have fallen by only about 1 °C – not such a dramatic climate shift.

This view is highly contentious. Alan Robock from Rutgers University in New Brunswick, New Jersey came up with the original simulations and has stood by them. Yet archaeological and geological work in India seems to support Graf's claims, suggesting that the environmental impact of the super-eruption was much less than previously imagined.

Firstly, had there been a sudden deforestation event caused by the cooling and drying of the atmosphere, topsoil no longer anchored by trees would be expected to wash down into valleys, where it would quickly accumulate. Yet when Peter Ditchfield of the University of Oxford looked for such an influx of soil, he could find no trace of it.

To build further evidence, Ditchfield analysed the ratio of different carbon isotopes – which are each absorbed at different rates by different plants – in ancient plant remains in the Jwalapuram region of southern India and the Middle Son river valley in central northern India, both of which are around 2,000 kilometres from Toba. He saw only a slight increase in the carbon-13 isotope after the Toba eruption, which suggests just a small increase in grassland environments at this time. In other words, woodlands were not obliterated by Toba.

Nevertheless, hominin species living at the time of the eruption would undoubtedly have faced tough conditions. The blanket of ash, for example, would have been quickly washed into the freshwater supplies: Ditchfield found deposits up to 3 metres deep on the valley floors where rivers would once have flowed. And there is no doubt that in the years immediately following the eruption the early humans would have had to adjust to colder temperatures, probably having to economize significantly as food resources dwindled. But just because life was difficult for humans after Toba does not mean that the situation was catastrophic.

Mike Petraglia at the Max Planck Institute for the Science of Human History in Jena, Germany, led a team to investigate a number of sites at Jwalapuram. One has been particularly fruitful. Labelled Jwalapuram 22, it was probably a hunter-gatherer camp. It has yielded more than 1,800 tools, including stone flakes, scrapers and points – the everyday tools for cutting and scraping – and the stone 'cores' left over following tool manufacture.

Surprisingly, hominin life appeared to continue in the same vein immediately after the eruption, with hundreds more stone tools in the layers immediately above the ash fall. The team uncovered a similar story 1,000 kilometres further north of Jwalapuram, in the Middle Son river valley. Again, that is not to say that the eruption was an easy ride for the hominins living in India. Jwalapuram and the Middle Son valley may have been special cases – refugia in which hominin populations sheltered when times got tough. Still, the findings present a challenge to the traditional view of Toba as a devastating catastrophe for hominins alive at the time.

The debate about what (if anything) the Toba super-eruption did to early humans still continues, but the idea that it was a catastrophe does not look well supported.

The impact of inbreeding

Another key factor shaping our evolution may have been rather inauspicious. It seems that, for thousands of years, our ancestors lived in small and isolated populations, leaving them severely inbred. The inbreeding may have caused a host of health problems, and it is likely that small populations were a barrier to the development of complex technologies.

From the sequenced genomes of Neanderthals and Denisovans, David Reich of Harvard Medical School in Boston, Massachusetts found that both species were severely inbred, due to their small populations, and had an extraordinarily low level of genetic diversity. This is in line with previous evidence of small populations. It has been estimated that, in the distant past, human populations were probably only in the thousands or at best tens of thousands.

Fossils suggest that the inbreeding took its toll, according to studies by Erik Trinkaus of Washington University in St Louis, Missouri. Those he has studied have a range of deformities, many of which are rare in modern humans. He thinks such deformities were once much more common.

Such small populations may have affected the course of culture and technology. Larger populations retain more knowledge and find ways to improve technologies. This 'cumulative culture' is unique to humans, but it could emerge only in reasonably large populations. In small populations knowledge is easily lost, which explains why skills like bone-working show up and then vanish.

Tiny populations may have prevented Neanderthals and Denisovans from developing cumulative culture, placing limits on their cultural complexity. The same thing held our species back, until the population reached a critical density, unleashing the power of culture – at which point there was no stopping us.

Cooperative breeding

One further influence on human evolution is our capacity for caring for other people's offspring. As we have seen, humans are an unusually cooperative species, and this may be key to our success. A handful of researchers have argued that cooperation arose from the development of a single behaviour: sharing childcare. They claim that the care, nutrition and protection of youngsters by adults other than the mother bring about profound psychological changes in a species. In humanity, this may have paved the way for the enhanced cooperation and altruism that led to culture, language and technology.

There is no doubt that humans are extraordinarily social when compared with most of the animal kingdom. We are generally good at reading other people's emotions and adapting our behaviour appropriately, we work well in teams on highly complex projects, and we sometimes even extend our kindness to perfect strangers. This capacity for cooperation is thought to have been essential for the development of culture and technology, making it one of the defining changes in our evolution. So where did it come from?

Chimps have a mean streak, but monkeys called marmosets are remarkably altruistic, much like humans. Carel van Schaik and Judith Burkart began to wonder what might explain these acts of altruism. One similarity seemed to stand out: humans and marmosets are 'cooperative breeders'. Much more than most other primates, the adults of a marmoset group willingly protect and actively feed one another's young, usually without any prompting. Chimps, by contrast, are independent breeders who will rarely help another's family. They are not even particularly giving to their own infants.

Van Schaik and Burkart now suspect that the evolution of cooperative breeding might have paved the way for greater altruism more generally.

10
Unanswered questions

This book is full of words, many of which are 'maybe', 'perhaps' and 'could'. While the science of human evolution has plenty of hard facts in the form of fossils, stone tools and DNA, the interpretations we place on them are all our own. Put simply, we are telling stories to fit those facts — and scientists do not agree about which stories are correct. This final chapter sets out the major areas of contention, as of summer 2017. There may well be new arguments by the time you read this book.

Rewriting the story

By the year 2000 a fairly complete story of human evolution seemed within reach. The majority view went something like this.

Roughly 5.5 to 6.5 million years ago, somewhere in an East African forest, there lived a chimpanzee-like ape. Some of its descendants remained in the forests, eventually evolving into modern chimps and bonobos. But some left the forests and moved on to the savannahs. They became our hominin lineage. The hominins adapted to their new environment, most obviously by evolving to walk on two legs. By about 4 million years ago they had given rise to a successful group called the australopiths.

About 2 million years ago some of these relatively ape-like australopiths went through a significant transition, gaining larger brains and long legs to become the earliest 'true' humans. One of these early humans – *Homo erectus* – used its relatively large brain and long legs to march out of Africa, becoming the first hominin to do so. African humans continued to evolve larger brains in an apparently inexorable fashion, with additional waves of bigger-brained humans migrating out of Africa over the next million years or so. One of these waves probably gave rise to the Neanderthals – which, by the year 2000, were generally viewed as a distinct evolutionary offshoot and not ancestral to living people.

It was the humans remaining in Africa who eventually evolved into our species, *Homo sapiens*. About 60,000 years ago *H. sapiens* also began moving out of Africa. It may have met the Neanderthals. It may even have played a part in their extinction. But because twentieth-century studies of Neanderthal fossils – and Neanderthal DNA – suggested that neither was *H. sapiens*-like, it was far from clear that the two species interbred.

In 2017, just 17 years later, almost all of these assumptions have been called into question. Our last common ancestor with chimps may not have been much like a chimp. The split with chimps might have occurred much earlier than we thought. Hominins may have become bipedal before leaving the trees, not after. The australopiths did not necessarily give rise to true humans as we once assumed – but, against expectations, they might have migrated out of Africa and evolved into Indonesian hobbits. Small-brained humans apparently survived alongside large-brained species, perhaps even our own. And our species apparently regarded the other ancient humans it met as physically – and perhaps behaviourally – similar enough to interbreed with them.

Did we really come out of Africa?

One of the most long-running controversies in all of anthropology concerns where our species first evolved.

One school of thought believes that the fossil evidence indicates that modern humans arose relatively recently in Africa. This species then spread throughout the Old World, replacing existing populations of earlier forms of *Homo*, including the Neanderthals. This is known as the 'Out of Africa' hypothesis, and nowadays it is the majority view. In most of the earlier chapters, we have largely taken it as read.

But there is the second school of thought, which argues that modern humans appeared more or less simultaneously in Africa, Europe and Asia, evolving *in situ* from an earlier hominid species, *Homo erectus*, which had migrated from Africa into much of the rest of the Old World about a million years ago. This is known as the 'multiregional hypothesis'. Among the most vocal champions of this theory have been Milford Wolpoff of the University of Michigan in Ann Arbor and Alan Thorne of the Australian National University in Canberra, who died in 2012.

The differences between the two hypotheses are great, both in the evolutionary model for the origin of modern *H. sapiens* and in the predictions they make about the nature of the fossil record.

For instance, according to the Out of Africa hypothesis, all modern human populations derive from an original African stock. Any anatomical features that might have evolved in different regions of the world in earlier *H. erectus* or other populations will have been lost. In other words, there will be no specific regional continuity of anatomical characteristics from, say, a million years ago through to the modern world.

By contrast, the multiregional model predicts continuity of regionally developed characteristics. For instance, the model argues that modern Chinese populations evolved ultimately from populations of *H. erectus* that entered China as much as a million years ago. These original populations evolved over time, gaining the characteristics of modern humans but retaining at least some of their original features. The same argument applies to populations elsewhere in the Old World.

For many years, the debate hinged on whether there was anatomical evidence of regional continuity in the fossil record. If the multiregional hypothesis is correct, then *H. erectus* fossils in, say, China should resemble modern Chinese people, while fossils from Africa should resemble modern African people. Proponents of the multiregional hypothesis claim to have found exactly this. They argue that ancient Chinese *H. erectus* fossils foreshadow the morphology of modern Chinese populations in features such as their relatively small, flat faces, and prominent and delicately built cheek bones.

However, the fossil evidence was always contested. Then, from the 1980s onwards, genetics entered the fray. This turned the tide in favour of the Out of Africa hypothesis. The key finding was that all modern humans seemed to be descended from the same, rather small population that lived around 150,000 years ago – in Africa. A series of studies found essentially this pattern and by the early 2000s many researchers considered the matter settled.

However, multiregionalists like Wolpoff have continued to fight their corner. They have pointed to other fossils, such as an Indonesian fossil known as 'Java Gal' that was described as being halfway between *H. erectus* and *H. sapiens* – seemingly implying that Indonesian *H. erectus* evolved into Indonesian *H. sapiens*.

Wolpoff also attempted to capitalize on the discovery that humans had interbred with other species, such as Neanderthals. He argued that the genetics has revealed several lineages of humans in the Pleistocene, all of which could interbreed, and that this was in line with the multiregional hypothesis.

Unsurprisingly, Out of Africa supporters disagree. They argue that the implication from the multiregional model was that Neanderthals gradually changed into modern humans. But that is not what the fossil record shows. Instead, it shows Neanderthals staying pretty much the same up until around 30,000 or 40,000 years ago, and then disappearing.

The multiregionalists are not giving up, but the tide of opinion is against them.

Why did we become bipedal?

Charles Darwin suggested that our ancestors first stood upright to free their hands for toolmaking. But we now know that cannot be right. Chimpanzees use tools and they are not bipedal.

The trouble with bipedalism is that proficient walking has many advantages, but acquiring the skill requires anatomical changes, and in the meantime you will be slow, clumsy and unstable. One possibility is that bipedalism began in the trees. After all, orang-utans and other primates walk upright along branches when feeding. This fits with what we know about the lifestyle of the first bipeds, but does not explain why they evolved specialist anatomy.

Some people believe it evolved to allow males to access more food, so that they could help feed their partners and offspring. But this idea presupposes a very early origin of monogamy, which the evidence does not support.

Another possibility is that individuals who could wander farther than others had access to a wider variety of food sources, allowing them to live longer and produce more surviving offspring. In addition, bipedalism would have left their hands free to carry things and, being taller, they may have been better at spotting predators.

All of this would have set the stage for a second phase of evolution around 1.7 million years ago, when our ancestors left the forests for the savannah. This is when the greatest anatomical changes took place, with shoulders pulled back, legs lengthened and a pelvis adapted to life on two legs.

There are many possible reasons why bipedalism took off at this point. Walking upright might have helped individuals deal with the scorching heat of the open grassland, allowing air to circulate around the body while minimizing direct exposure to the sun. It would also have increased mobility.

The Out of Asia hypothesis

The Out of Africa hypothesis has been challenged in another way in the twenty-first century. Here, the argument is not about the origins of *Homo sapiens* but the entire *Homo* genus: an earlier stage in our evolution.

Some prominent researchers have come round to the idea that hominins may have left their African cradle much earlier than we thought and undergone critical evolutionary transitions further north. There are even claims that the appearance of our genus *Homo* may have occurred under Eurasian rather than African skies. The catalyst behind this radical rethink is that problematic little hobbit, *Homo floresiensis*.

From the beginning, the hobbit did not fit the standard picture of human evolution. Some of the remains found on the Indonesian island of Flores were thought to be just 18,000 years old, implying that the hobbit was alive at least 10,000 years after every other hominin except our own species had become extinct. As we have seen, this has now been disproved (see Chapter 5) and the remains are now thought to be more like 50,000 years old, but that is still recent for such a small-brained creature. The cranial volume of the single hobbit skull found so far is around 420 cubic centimetres, about one-third the size of a modern human's. Yet stone tools found with hobbit bones suggest that the hominin was capable of sophisticated behaviour.

There are so many primitive features in the hobbit skeleton that its discoverers began to talk seriously about *H. floresiensis* being derived from something even more primitive than *H. erectus* (see Chapter 5). Somewhere towards the very top of the list of the hobbit's likely recent ancestors is an australopith.

This idea borders on the incendiary. Conventional wisdom suggests that the australopiths evolved in Africa about 4 million years ago and died out there 2.8 million years later, without ever having left. Perhaps it was their short legs that discouraged them from making the long trek out of Africa. Certainly, it was not until the taller members of our own genus appeared, towards the end of the australopith age, that hominins began to explore the wider world.

The hobbit remains hint at an alternative. Perhaps an australopith did manage to escape Africa before the *Homo* genus evolved, and perhaps it survived long enough in Eurasia to evolve into the hobbit. If so, we should perhaps have found some fossil evidence for these ancient Eurasian australopiths by now. However, the environmental conditions in East and South Africa favoured preservation of human fossils in a way that conditions across Asia did not.

Still, there is one site in Eurasia that could fit with the idea that australopith-like hominins made it out of Africa. There are also hints that these enigmatic Eurasian australopiths did more than evolve into the hobbits found on Flores: they may have given rise to our own genus.

In 1991 researchers excavating the medieval town of Dmanisi, Georgia, in the Caucasus came across the earliest hominin remains so far found outside Africa. There is still some debate over exactly where the 1.77-million-year-old Dmanisi hominins fit in the human evolutionary tree, but most would classify them as *H. erectus*. Their age and primitive features suggest that they were among the earliest members of this species, implying that *H. erectus* wasted little time in leaving the East African region in which it first appeared perhaps 1.87 million years ago. Conventional thinking has it that this was the first time a hominin ventured out of Africa, with Dmanisi offering a unique snapshot into the very moment that humans went global.

Then, in 2011, came surprising news from Dmanisi that challenged this picture. Continued excavations had found evidence that the Georgian site was first occupied at least 1.85 million years ago – essentially at the same time that *H. erectus* appeared in East Africa. This suggested to some that *H. erectus* could have evolved in Eurasia. If so, the fossils at Dmanisi are not a snapshot of the first hominin migration north out of Africa, but rather an indication of *H. erectus* in the act of migrating south into the land of its forefathers.

More broadly, the new dates of occupation at Dmanisi mean that *H. erectus* could have evolved from an australopith that left Africa around or before the 2-million-year mark. This is crucial, because *H. erectus* is often seen as the direct ancestor of our species. So if it evolved in Eurasia before moving into Africa where our species evolved about 350,000–200,000 years ago, modern humanity is arguably the product of both an African and a Eurasian cradle (see Figure 10.1).

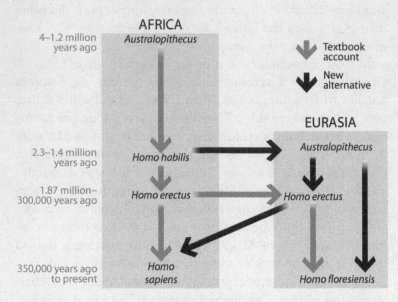

FIGURE 10.1 We used to think our genus *Homo* evolved in Africa but recent finds suggest that our ancestors may have taken a detour to Eurasia.

It should be stressed that the evidence from Flores and Dmanisi is merely compatible with these radical new ideas rather than strongly supportive of them. The fossil evidence from Eurasia is still meagre, so there is no hard evidence that australopiths migrated out of Africa.

Why was technological development so slow?

Sharp stone flakes found in the 1990s in a parched riverbed in the Afar region of Ethiopia are the oldest tools yet discovered. They date from 2.6 million years ago – although there is indirect evidence of tool use at 3.4 million years ago. Either way, it would be at least another million years before our ancestors made their next technological breakthrough.

Then, instead of using the chips off a river cobble as blades and scrapers, someone realized that the cobble itself could be worked into a tool: a rough hand axe. Another million years passed before early modern humans perfected this technique. What took them so long?

Intelligence must have played a part. In the 2 million years after the appearance of the first tools, hominin brain size more than doubled, to around 900 cubic centimetres. So, although tools appear not to have developed much, their production is underpinned by great cognitive advance.

In his 2011 book *The Origin of Our Species*, Chris Stringer of the Natural History Museum in London identified another reason – demography. Modern humans have large populations, with many people copying and many ways to pass on information. Our long lives also permit transfer of ideas down the generations, whereas *Homo erectus* and *Homo heidelbergensis* probably had a maximum lifespan of around 30 years, and Neanderthals maybe 40. They had to grow up very fast and there was much less networking between groups.

Furthermore, our ancestors may have shunned change, since life would have been challenging enough without risky experimentation. In his 2011 book *Wired for Culture*, Mark Pagel at the University of Reading, UK argues that hominins before *Homo sapiens* did not have what it takes to innovate and exchange ideas, even if they wanted to. He draws a comparison with chimps, which can make crude stone tools but lack technological progress. They mostly learn by trial and error, whereas we learn by watching each other, and we know when something is worth copying. If Pagel is correct, then social learning is the spark that ignited a technological revolution (see Chapter 8).

Redrawing our family tree

One of the most iconic scientific illustrations of all time is 'The March of Progress', drawn for a popular science book in 1965. The image lined up all the early relatives of humans known at the time in chronological order (see Figure 10.2). The artist, Rudolph Zallinger, sketched them striding purposefully across the page, seemingly becoming more advanced with each step. It gave the impression – despite the book saying otherwise – that human evolution was a linear progression, from small-brained tree climbers to bipedal big-brained modern humans.

This much-copied image has been criticized for its oversimplification, but until recently our evolutionary past was not actually thought to be a great deal more complex, give or take the odd dead-end side-shoot. But, as of the twenty-first century, anthropologists are having a rethink. In particular, it looks if as many species of human-like apes were around during the crucial period from 2.5 to 1.8 million years ago, when the first upright apes with relatively large brains evolved. What's more,

FIGURE 10.2 'The March of Progress', a simplified vision of human evolution

the East African hominin long seen as our direct ancestor may be just a cousin, with our true roots lying elsewhere. Our family tree may have to be completely redrawn.

The story once looked so straightforward. A hominin found in Tanzania's Olduvai Gorge in 1960 neatly explained how *Australopithecus* became *Homo*. The fragmentary remains belong to a species with a brain roughly 50 per cent larger than the average australopith but half the size of our own brains. It first appeared around 2.3 million years ago, just as most australopiths were vanishing but before *Homo erectus* had evolved. And the East African region it lived in had formerly been home to small-brained australopiths, and would later be inhabited by *H. erectus*.

Its discoverers opted to make this hominin the first member of our genus, naming it *Homo habilis*. For 50 years *H. habilis* was central to the story of our origins. It was the right hominin in the right place at the right time to be our direct ancestor.

It was not the only hominin around in this place at this time. In the years before *H. habilis* was discovered, another australopith now known as *Paranthropus boisei* had been found. *P. boisei* has some unusual, almost gorilla-like features, though, and is clearly a side branch rather than a direct human ancestor. The evidence, then, seemed to point to a simple picture not wildly different from 'The March of Progress'. But others think the picture is more complicated.

In 1972 a research team working in the Koobi Fora region of northern Kenya found a skull from the time that *H. habilis* was alive, with a brain slightly larger and a face considerably broader and flatter – that is, with prominent cheekbones – than that of any known specimen of *H. habilis*. One researcher called it *Homo rudolfensis*. But Meave Leakey, based at the Turkana Basin Institute in Nairobi, Kenya, and a member of the team that found the skull, prefers to stick to specimen numbers. The skull's number is KNM-ER 1470, or 1470 for short.

For almost 40 years, the 1470 skull remained an anomaly. That changed in 2007, when Leakey and her colleagues began finding similar fossils in the same region. The discoveries, which include a new skull fragment with the flat facial features of the 1470 skull, were revealed in 2012. The new find belonged to a juvenile, not an adult. This suggests that 1470 was no anomaly, but instead belonged to a distinct species that was born and grew up with a flat face. Although in some respects that flat face is rather like ours, it seems to be far too broad to be one of our direct ancestors.

With *H. habilis* and *P. boisei*, this means that at least three species of hominins were living in East Africa around 2 million years ago. Or were there more? A scrap of skull from Koobi Fora that dates back 2 million years is strikingly reminiscent of *H. erectus*, raising the tantalizing possibility that it lived in East Africa at the same time as *H. habilis* and the 1470 lineage.

Four hominin species living at the same time would be exceptional. And yet Bernard Wood at George Washington University in Washington, DC has suggested there was yet another hominin, known from a lone jawbone in the Koobi Fora collection.

Potentially, then, East Africa was home to five species of hominin just as our genus was finding its feet. How they are all related is far from clear, but *H. habilis* still looked like the most likely direct human ancestor until an unexpected discovery thousands of kilometres to the south of Koobi Fora and Olduvai Gorge. That was Lee Berger's 2010 discovery in South Africa of *Australopithecus sediba* (see Figure 10.3 and Chapter 3).

Along with the East African hominins, this means that there were as many as six species living in Africa 2 million years ago – a level of diversity unprecedented in 7 million years of hominin evolution. *A. sediba* is arguably the most surprising of the six.

In some ways, such as brain size, *A. sediba* resembles other australopiths. But what makes it a strange ape is that in other ways it resembles humans. The more Berger looked at the skeletons, the more convinced he became that *A. sediba* is a pivotal species in our ancestry. He thinks that the characteristics *A. sediba* possesses, including small *Homo*-like teeth and a tapered *Homo*-like waist, put it on the lineage leading to *H. erectus*.

But *A. sediba*, as critics are quick to point out, is everything that *H. habilis* is not: it is a small-brained australopith living in southern Africa 2 million years ago – a good 300,000 years after the larger-brained *H. habilis* first appeared in East Africa. They say *A. sediba* is the wrong hominin in the wrong place at the wrong time to be our direct ancestor. It is just too young to lead to *Homo*, is the argument.

→ Key relationship ⋯⋯> Possible relationship

One view is that our family tree is very simple.

Others think that there were many different species, making it more complex.

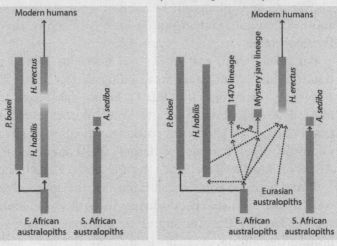

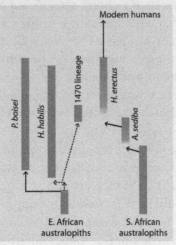

Most radical of all is the idea that modern humans arose from a South African ape.

FIGURE 10.3 We now have at least three competing visions of our family tree based on fossils found so far.

Berger has a simple answer to this criticism: *H. habilis*, the oldest member of our genus, is not one of our direct ancestors. Its relatively large human-like brain gives the impression that it is, but appearances can be deceptive. For Berger, *A. sediba* is a better candidate for the origin of *erectus* than *habilis* ever was. Its hand, dentition and what we can see of its skull morphology – other than the cranial capacity – are more like those of *H. erectus*.

As of 2017 there is no consensus on whether *H. habilis* or an *A. sediba*-like australopith was our direct ancestor.

Are other hominins still alive?

Legends of human-like creatures, such as Bigfoot and the Yeti, have entranced people for centuries. They make for good stories, but could there be any truth in them?

It seems unlikely. In a 2009 study Jeff Lozier at the University of Alabama in Tuscaloosa examined the location of all Bigfoot, or Sasquatch, sightings. He found that these 'haunts' are identical to those of the black bear, suggesting it could simply be a case of mistaken identity. Similarly, David Coltman at the University of Alberta in Edmonton, Canada once analysed a tuft of hair from a supposed Bigfoot, and found that it came from a bison. Of course, new species of primate are occasionally found in remote regions, so there is a slim chance that there may be something out there.

Nevertheless, a few scientists are willing to contemplate the idea that *Homo sapiens* is not alone. After all, other hominin species coexisted alongside our ancestors for most of human history. And our family tree can still surprise us, as happened with the discovery of the pint-sized hobbits in 2003.

At this point there is no hard evidence for the existence of another living hominin species, and it may well be that none exists. But it would be foolish to completely discount the possibility.

Are we still evolving?

Many biologists would like to believe that significant human biological evolution stopped between 50,000 and 100,000 years ago, before the races diverged, because this would ensure that racial and ethnic groups are biologically equivalent. But recent discoveries suggest that we must reject the idea that human evolution stopped dead 50,000 years ago or more. There is every reason to believe that it is going on right now.

Take the discovery in 2005, by Bruce Lahn of the University of Chicago, of two genes involved in brain development that emerged in recent human history and swept quickly through the population. One, a version of a gene called *microcephalin*, arose between 14,000 and 60,000 years ago and is now carried by 70 per cent of people. The other, a variant of the *ASPM* gene, is as recent as 500 to 14,000 years old and is now carried by about a quarter of the global population.

The discovery of ongoing human evolution raises many questions, some of them uncomfortable. What if racial groups turn out not to be biologically equivalent? Is natural selection still a driving force in humans, given that our survival is often less dependent on genes than on technology? To what extent might a changing genome lead to changes in attributes we value, such as intelligence? What might our species look like a thousand years from now? Contemporary human evolution is a minefield.

In the loosest sense of the word, evolution is simply the change over time in a species' gene pool – all the genes in all the individuals alive at one time. In that sense, all species are evolving, even those that reproduce by cloning, because DNA inevitably changes over time through random mutation, and because some individuals of a species will have more offspring than others. But are any selection pressures at work – like the need to reach into tall trees that prompted giraffes to evolve longer necks?

Steve Jones, a geneticist at University College London, has argued that natural selection is no longer important for humans. He points out that natural selection works by ensuring that individuals whose genes are best adapted to the prevailing environment are most likely to survive and reproduce. But in the developed world, survival arguably no longer depends on genes. Almost all babies now live to adulthood, whereas in past centuries only half of them did.

There is also a more level playing field in the reproduction game: in the Middle Ages a few rich people were survived by many children while many of those in poverty were not. Jones has calculated that the changes in survival and reproduction rates have led to a decrease of around 70 per cent in the opportunity for natural selection to act today, compared with the time when our ancestors lived as peasant farmers.

Still, that is not quite 'zero' natural selection. Clearly, genes can still make a difference to survival and reproduction. One obvious example is genes that confer resistance to emerging diseases. Some parts of Africa, for example, have seen an increase in the frequency of a gene called $CCR5\text{-}\Delta32$, which offers some protection against infection with HIV-1. There are other, more puzzling examples. One form of the dopamine receptor gene *DRD4* has become much more common over the past few thousand years. The rate of increase suggests that the gene has been positively selected for, though it is not clear why: the variant is associated with attention deficit hyperactivity disorder.

In fact, a study published in 2006 identified human genes that have been selected for in the past 10,000 years: not just one or two genes, but more than 700. So natural selection is still at work, and some evolutionary biologists believe that it should come as no surprise. They point out that we live in an era of rapid technological progress, and hence a fast-changing environment, exactly the conditions under which you would expect natural selection to act. Technological change has clearly driven natural selection in the past. The invention of dairy herding, for example, selected for a gene that gives adults the ability to digest milk sugars. So why not now?

Some experts argue that technological change does not necessarily drive natural selection. Once culture emerged, they say, it provided non-genetic means to adapt to change, such as more technology or culturally inherited changes in behaviour. Although that is true in many ways, it does not necessarily mean that evolution has stopped. Technology and medicine, by enabling almost everyone to have children, could be causing 'reverse evolution' by preventing unfit genes from being purged from the gene pool. Relaxed selection, combined with a high mutation rate, could potentially be causing gradual deterioration of many functions, especially disease defences.

There are also plausible ways in which culture itself could be driving natural selection, according to Christopher Wills of the University of California, San Diego. In his 1993 book *The Runaway Brain*, he argues that there has been, and still is, positive feedback between our culture and our genes that led to the rapid evolution of the most characteristic human attribute, the mind. It began when the relatively advanced brains of our ancestors allowed them to succeed because of their wits rather than physical attributes. This is one reason why Lahn's discovery of recent brain evolution created such a stir.

Natural selection, however, is not the only reason why a gene might become more prevalent. It is also possible that the driving force is sexual selection. Among the most prominent supporters of this idea is Geoffrey Miller of the University of New Mexico, Albuquerque, author of *The Mating Mind* (2001). He believes that the rate of human evolution is accelerating, and that selection for sexually desirable traits is the driving force. Our species is experiencing high levels of migration, outbreeding and cross-ethnic mating, which, he argues, are recombining our genes at unprecedented rates.

In the future, parents may try to eliminate traits that they personally find undesirable, but it is impossible to predict how that will affect the human gene pool. Our genes may also get some shiny new high-tech additions, as humans merge with technology to become cyborgs and biological evolution is rendered obsolete.

Finally, if we colonize other planets, the colonists – and the animals and plants they take with them – will undergo dramatic evolutionary changes as they adapt to their new conditions. It is possible that colonists would even become a separate species, if there was no interbreeding with people on Earth.

All in all, it is hard not to conclude that humans are still evolving, probably quite rapidly. Wherever we end up, it seems clear that the story of human evolution has only just started.

Conclusion

The most important point to make about the story of human evolution is that we have not yet written it. We thought we had: by the late 1990s anthropologists were converging on a story. But the flurry of new discoveries since 2000 – not just from fossils but from DNA evidence, too – has thrown the story into confusion.

It is too soon to tell how all the new pieces of the puzzle will fit together. But here is an interesting speculation to end on: what if the history of palaeoanthropology had been reversed? In other words, suppose *Homo naledi* and *Australopithecus sediba* – the two species discovered most recently – had in fact been discovered decades ago, with 'classic' species like *Homo erectus* and the Neanderthals being uncovered only in the last decade.

If all we had had to go on for decades were *H. naledi* and *A. sediba*, it seems likely that our ideas about human evolution would have been very different. We would surely have assumed that our species originated in southern Africa, rather than East Africa as was long thought. Faced with 'new and strange' discoveries like *H. erectus*, we would find ourselves thoroughly confused.

It seems virtually certain that there are more species to be found – perhaps many more. DNA will continue to reveal more about how and when our species' traits appeared. And the picture will keep changing. Unlike some branches of science, human evolution has not yet figured out its central message.

Fifty ideas

This section offers ideas for how to explore the subject of human evolution in greater depth.

Ten places to visit

1 **Atapuerca Mountains, Spain.** This archaeological site is a UNESCO World Heritage Site, noted for the large numbers of *Homo heidelbergensis* remains. More information at www.atapuerca.org/

2 **Dmanisi Museum-Reserve, Georgia.** This site has yielded some of the oldest hominin remains outside Africa. The museum is open seasonally. Details at http://museum.ge/index.php?lang_id=ENG&sec_id=51

3 **Lake Turkana, Kenya.** This rich hominin fossil site is covered by three national parks: Sibiloi, South Island and Central Island. Sibiloi contains the archaeological sites. More information at www.kws.go.ke/national-parks

4 **Cradle of Humankind, South Africa.** This sprawling series of limestone caves has yielded many crucial fossils, including the first specimens of *Australopithecus sediba*. There is a visitor centre and exhibition called Maropeng. More information at www.maropeng.co.za/

5 **Neanderthal Museum, Germany.** Located in the region where the Neanderthals were first identified from fossils, this museum recounts the story of humanity. Visitor information at www.neanderthal.de

6 **Smithsonian National Museum of Natural History, Washington, DC, USA.** The Hall of Human Origins contains a wealth of information. More information at https://naturalhistory.si.edu

7 **Zhoukoudian Caves, China.** At this site, archaeologists discovered the famous Peking Man fossils – most

of which were sadly lost during the Second World War. There is a museum on the site. Details at www.china. org.cn/english/MATERIAL/31256.htm

8 **Ngorongoro Conservation Area, Tanzania.** This World Heritage Site encompasses Oldupai (previously Olduvai) Gorge, where many hominin fossils have been found, and the famous preserved *Australopithecus* footprints at Laetoli. More information at www. ngorongorocrater.org

9 **Naturalis Biodiversity Centre, Netherlands.** This museum houses many crucial fossils, including some of the first *Homo erectus*. It is closed for refurbishments until the end of 2018. Details at www.naturalis.nl

10 **Blackwater Draw National Historic Landmark and Museum, New Mexico, USA.** This museum is near the site of the first Clovis excavation, which purported to reveal how the Americas were first colonized. Information at www.bwdarchaeology.com/

Ten quotes

1 'Among the multitude of animals which scamper, fly, burrow and swim around us, man is the only one who is not locked into his environment.' (Jacob Bronowski, *The Ascent of Man*, 1973)

2 'If it be an advantage to man to have his hands and arms free and to stand firmly on his feet, of which there can be no doubt from his pre-eminent success in the battle of life, then I can see no reason why it should not have been advantageous to the progenitors of man to have become more and more erect or bipedal.' (Charles Darwin, *The Descent of Man*, 1871)

3 'Doing palaeoanthropology is much like doing a jigsaw puzzle without a picture of the pattern.' (William H. Calvin, *A Brain for all Seasons: Human Evolution and Abrupt Climate Change*, 2002)

4 'With the throwaway line, "Why don't you call her Lucy?" came total commitment from everyone on the team by breakfast the next day. "When are we going back to the Lucy site?" people asked. "How old do you think Lucy was when she died?" Immediately she became a person.' (Donald Johanson, discoverer of the 'Lucy' fossil, interviewed by *Scientific American* in 2014)

5 'I sometimes try to imagine what would have happened if we'd known the bonobo first and the chimpanzee only later – or not at all. The discussion about human evolution might not revolve as much around violence, warfare and male dominance, but rather around sexuality,

empathy, caring and cooperation.' (Frans de Waal, *Our Inner Ape: The Best and Worst of Human Nature*, 2005)

6 'Being somewhat of a curmudgeon, what does an odd population of mini-hominins on an isolated island really tell us about us? You could almost say, so what?' (Anthropologist Richard Leakey on the 'hobbit' *Homo floresiensis*, interviewed by *New Scientist* in 2009)

7 'We may feel humbled by the survival of our species, though set about by vicissitudes, and we may marvel at our ancestors' ingenuity and adaptability – but we must remember ... that they were just *people* – like you and me.' (Alice Roberts, *The Incredible Human Journey*, 2010)

8 'Anyone who has experienced first hand the overwhelming power of the life-sized painted bulls and horses in the Lascaux Cave of south-western France will understand at once that their creators must have been as modern in their minds as they were in their skeletons.' (Jared Diamond, *Guns, Germs and Steel,* 1997)

9 'Believe it or not – and I know that most people do not – today we may be living in the most peaceable era in our species' existence.' (Steven Pinker, *The Better Angels of Our Nature*, 2011)

10 'If I thought of man as the final image of God, I should not know what to think of God. But when I consider that our ancestors, at a time fairly recent in relation to the earth's history, were perfectly ordinary apes, closely related to chimpanzees, I see a glimmer of hope. It does not require very great optimism to assume that from us human beings something better and higher may evolve.' (Konrad Lorenz, *On Aggression*, 1963)

Ten questions to ponder

1 Why are we the only hominin species still alive?

2 Did Neanderthals have religious beliefs?

3 Why did Stone Age people paint on cave walls?

4 Why did we lose almost all our body hair?

5 Would you have sex with a Neanderthal? What about a *Homo erectus*?

6 Is there anything genuinely unique about us as a species?

7 Why did our ancestors become farmers and city-dwellers?

8 Could Lucy the *Australopithecus* speak?

9 If a Neanderthal got dressed in human clothes and went for a walk, would people in the street realize who he/she was?

10 How many hominin species are left for us to find?

Twenty key discoveries

1 **1829:** Fragments of a human skull are found in a cave in present-day Belgium. The 'Engis 2' skull is the first ever hominin fossil found. It is now known to belong to a Neanderthal.

2 **1856:** The first Neanderthal fossil to be identified as such, Neanderthal 1, is discovered in the Neander Valley in Germany.

3 **1891:** Eugène Dubois discovers the first *Homo erectus* on the island of Java. He calls it 'Java Man'.

4 **1907:** The first *Homo heidelbergensis* is found in a sand-pit in Germany.

5 **1909:** A gold prospector finds a fossil primate, later known as *Proconsul* and dated to about 24 million years ago.

6 **1924:** The Taung Child, the first fossil of an *Australopithecus*, is found in South Africa. It is described by Raymond Dart the following year.

7 **1959:** Mary Leakey finds a cranium called 'Nutcracker Man'. After a protracted debate, it is decided that it belongs to *Paranthropus boisei*.

8 **1960:** Jonathan and Mary Leakey discover 'Johnny's Child', the type specimen of *Homo habilis*.

9 **1964:** Excavations begin in the Atapuerca Mountains in Spain, home of Sima de los Huesos (the 'Pit of Bones').

10 **1971:** At Lake Turkana in Kenya, Richard Leakey discovers a lower jaw, which he identifies as belonging to *Homo ergaster*.

11 **1974:** Lucy, a 3.2-million-year-old *Australopithecus afarensis*, is discovered by Donald Johanson and colleagues in Ethiopia.

12 **1984:** Kamoya Kimeu discovers 'Turkana Boy', the most complete hominin skeleton ever found. It is thought to be a *Homo erectus*.

13 **1994:** Tim White describes *Ardipithecus ramidus*, discovered in Ethiopia.

14 **2000:** Brigitte Senut discovers *Orrorin tugenensis*, one of the oldest known hominins, in Kenya.

15 **2001:** A team of archaeologists discovers *Sahelanthropus tchadensis*, another ancient hominin, in Chad.

16 **2003:** *Homo floresiensis* is discovered on the island of Flores, Indonesia. Because of its diminutive stature, it is nicknamed the 'hobbit'.

17 **2008:** The first *Australopithecus sediba* is found by Lee Berger's son in South Africa.

18 **2010:** Svante Pääbo announces the discovery of the Denisovans, identified by genetic analysis of a lone finger bone found in Denisova Cave in Siberia.

19 **2013:** Two cavers discover fossils of *Homo naledi* in the Rising Star cave system in South Africa. The species is later described by Lee Berger.

20 **2017:** The oldest known fossils of our species are found in Morocco. They are 350,000 years old.

Glossary

Archicebus achilles An early primate that lived in Asia 55 million years ago

Ardipithecus kadabba An earlier species of *Ardipithecus* that lived 5.6 million years ago

Ardipithecus ramidus A hominin that lived in Africa 4.4 million years ago and may have been bipedal. The fossil 'Ardi' belongs to this species.

Australopithecus afarensis A bipedal hominin that lived 3 million years ago. The famous 'Lucy' fossil belongs to this species.

Australopithecus africanus The first *Australopithecus* to be discovered

Australopithecus sediba A bipedal hominin discovered in a cave in South Africa, which lived 2 million years ago

Bonobo A great ape closely related to chimpanzees, but which lives in a more peaceful and cooperative fashion

Chimpanzee A great ape and the closest living relative of humans

Denisovans A species of hominin that lived in Asia, and which is known only from fragmentary fossil remains that have yielded DNA

Evolution The process by which species gradually change and new species are formed

Gene A unit of genetic information, carried by the DNA molecules inside all living cells

Hominid Any member of the taxonomic group that includes humans, great apes (chimpanzees, bonobos, gorillas and orang-utans) and our extinct human-like relatives (*Australopithecus*, Neanderthals, etc.)

Hominin Any member of the smaller taxonomic group that includes modern humans and our extinct human-like relatives, but *not* the great apes. All hominins are also hominids, but not all hominids are hominins. The meanings of the words 'hominin' and 'hominid' have shifted in recent years, and older texts may define them differently.

Homo erectus A bipedal hominin that lived around 1.8 million years ago – the earliest hominin confirmed to have left Africa

Homo ergaster One of the earliest members of the *Homo* genus, preceding *H. erectus*. According to some interpretations, just a variant of *H. erectus*

Homo floresiensis An unusually small hominin species known only from the island of Flores, Indonesia; also known as 'hobbits'

Homo habilis Possibly the earliest *Homo* species

Homo heidelbergensis A wide-ranging hominin thought to be descended from earlier *Homo* species like *H. erectus*, and a possible ancestor of later Neanderthals and Denisovans

Homo naledi A hominin known only from a cave in South Africa, and which, despite being primitive in appearance, survived until just 236,000 years ago

Homo neanderthalensis A hominin species, commonly known as Neanderthals, that roamed Europe and Asia until roughly 40,000 years ago

Homo sapiens Our species, also known as 'anatomically modern humans'

Kenyanthropus platyops A hominin that lived in East Africa, apparently at the same time as several *Australopithecus* species

Mutation A change in a gene or other genetic material, which can be inherited by a creature's offspring and thus lead to evolutionary change

Orrorin tugenensis An early, ape-like hominin that lived around 6 million years ago

Paranthropus boisei A bipedal hominin that lived around 3 million years ago, alongside the australopithecines

Sahelanthropus tchadensis A very early hominin, dated to 7 million years ago and possibly close to the split between the hominin and chimpanzee lineages

Picture credits

All images © *New Scientist* except for the following:

Figure 1.2: Moment/Getty

Figure 2.1: Raul Martin/MSF/SPL

Figure 3.1: Michael Stravato/AP/REX/Shutterstock

Figure 4.1: Philipp Gunz, Simon Neubauer & Fred Spoor

Figure 4.2: Xinhua News Agency/REX/Shutterstock

Figure 5.1: Alfredo Dagli Orti/REX/Shutterstock

Figure 5.3: Lammel/ullstein bild via Getty

Figure 6.1: Newscom/Alamy Stock Photo

Figure 7.2: Vincent J. Musi/NGS/Getty

Figure 8.1: Glasshouse Images/REX/Shutterstock

Figure 9.2: Nigel Pavitt/AWL/Getty

Figure 10.2: DEA Picture Library/De Agostini/Getty

Index

Interested in learning more?

Learn more about the world and the big issues affecting us by downloading
New Scientist Instant Expert audiobooks and ebooks today.
All of the New Scientist Instant Expert audiobooks and ebooks in this
series are available to purchase from the Instant Expert app and from
https://instantexpert.johnmurraylearning.com/
Use **NSIE40** at https://instantexpert.johnmurraylearning.com/
for 40% off any purchase.